Earth's Core and Lower Mantle

The fluid mechanics of astrophysics and geophysics

A series edited by Andrew Soward
University of Exeter, UK and

Michael Ghil
University of California, Los Angeles, USA

Founding Editor: Paul Roberts
University of California, Los Angeles, USA

This book is part of a series. The publisher will accept continuation orders which may be cancelled at any time and which provide for automatic billing and shipping of each title in the series upon publication. Please write for details.

Earth's Core and Lower Mantle

Edited by
Christopher A. Jones, Andrew M. Soward
and Keke Zhang

Contributions from SEDI 2000, The 7th Symposium
Study of the Earth's Deep Interior Exeter,
30th July–4th August 2000

Taylor & Francis
Taylor & Francis Group

LONDON AND NEW YORK

First published 2003
by Taylor & Francis
11 New Fetter Lane, London EC4P 4EE

Simultaneously published in the USA and Canada
by Taylor & Francis
29 West 35th Street, New York, NY 10001

Taylor & Francis is an imprint of the Taylor & Francis Group

© 2003 Taylor & Francis

Typeset in Times New Roman by
Newgen Imaging Systems (P) Ltd, Chennai, India
Printed and bound in Great Britain by
TJ International Ltd, Padstow, Cornwall

Every effort has been made to ensure that the advice and information
in this book is true and accurate at the time of going to press. However,
neither the publisher nor the authors can accept any legal responsibility
or liability for any errors or omissions that may be made. In the case of
drug administration, any medical procedure or the use of technical
equipment mentioned within this book, you are strongly advised to
consult the manufacturer's guidelines.

British Library Cataloguing in Publication Data
A catalogue record for this book is available
from the British Library

Library of Congress Cataloging in Publication Data
A catalog record for this book has been requested

ISBN 0–415–30936–0

081105-11616 G2

Contents

Contributors

Keith D. Aldridge
Department of Earth and Atmospheric
  Science, York University
4700 Keele Street, Toronto, Ontario
Canada, M3J 1P3
E-mail: keith@yorku.ca

Friedrich H. Busse
Physikalisches Institut
Universität Bayreuth
Lehrstuhl Theoret Physik IV
Postfach 3008
D-95440 Bayreuth, Germany
E-mail: Friedrich.Busse@uni-bayreuth.de

Arthur R. Calderwood
Physics Department, University of
  Nevada-Las Vegas
Las Vegas, NV 89154, USA

Catherine G. Constable
Institute of Geophysics and Planetary
  Physics, Scripps Institution of
  Oceanography
University of California at San Diego
La Jolla, CA 92093-0225, USA
E-mail: cconstable@ucsd.edu

Eike Grote
Institute of Physics, University of Bayreuth
D-95440 Bayreuth, Germany

Dominique Jault
Laboratoire de Géophysique Interne et
  Tectonophysique – LGIT
Université Joseph-Fourier de Grenoble

BP 53, 38041 Grenoble Cedex 9, France
E-mail: Dominique.Jault@ujf-grenoble.fr

Christopher A. Jones
School of Mathematical Sciences
University of Exeter, Exeter
EX4 4QE, UK

Henri-Claude Nataf
Laboratoire de Géophysique Interne et
  Tectonophysique – LGIT
Université Joseph-Fourier de Grenoble
BP 53, 38041 Grenoble Cedex 9, France
E-mail: Henri-Claude.Nataf@ujf-grenoble.fr

Peter Olson
Johns Hopkins University, Olin Hall 3400
North Charles Street, Baltimore
MD 21209, USA
E-mail: olson@gibbs.eps.jhu.edu

Paul H. Roberts
Institute of Geophysics and Planetary
  Physics, UCLA, 405 Hilgard Avenue
Los Angeles, CA 90095, USA
E-mail: roberts@math.ucla.edu

Gerald Schubert
Department of Earth and Space Sciences
UCLA, Los Angeles, CA 90095-1567, USA
E-mail: schubert@ucla.edu

Radostin Simitev
Institute of Physics, University of Bayreuth
D-95440 Bayreuth, Germany
E-mail: radostin.simitev@uni-bayreuth.de

Preface

The study of the Earth's deep interior may be loosely defined as the study of those parts of the Earth that lie more than a thousand kilometres below the surface. These regions are obscured from direct view, but nevertheless a wealth of information about the deep interior has accumulated. Seismology, geomagnetism, the monitoring of the Earth's rotation, and accurate measurements of the gravity field have all played a role in this development. This observational knowledge has been supplemented by experiments, some conducted in extreme conditions of pressure and temperature. Theoreticians are also involved in unravelling the mysteries of the Earth's deep interior, both in building numerical models and in studying high-pressure physics. All this activity has brought together a multidisciplinary team of scientists, which was formalized into the SEDI (Study of the Earth's Deep Interior) organization in 1987.

The SEDI organization holds five-day symposia in alternate years. The 7th symposium was held in Exeter in the year 2000, 400 years after the publication of William Gilbert's book on the Earth's magnetic field, *De Magnete*. The symposium contained nine half day sessions on: (i) The structure of the inner core and the CMB; (ii) Seismic observation of the lower mantle; (iii) Core–Mantle–Inner core coupling; (iv) High-pressure physics of the Earth; (v) Experiments and short timescale phenomena; (vi) Paleomagnetic data; (vii) Dynamo models, theory and reversals; (viii) Secular variation, theory and the dynamo and (ix) Mantle convection. Abstracts of all the talks given appear in the Conference Programme and Abstract book. Many of the contributed lectures are published in a Special Issue of the journal *Physics of the Earth and Planetary Interiors*, **128**, pp. 1–244 (2001). This book contains eight chapters, which broadly span the main topics of the Symposium. They are written by established authorities in their fields, and are based on the invited lectures presented at this 7th SEDI symposium.

The 4th SEDI Symposium was held at Whistler Mountain, Canada, in 1994. The proceedings of that meeting were edited into the book *Earth's Deep Interior*. The subject area has moved on significantly in the last 6 years and the present book reflects the new developments. On the observational side, seismology has given us a new understanding of the nature of the inner core, revealing its anisotropy. Mapping this anisotropy, and trying to understand its origin, has become a major activity in the SEDI community. It has also given rise to a number of intriguing studies measuring the rate of inner core rotation relative to the mantle, though no consensus as to the magnitude of this effect has yet emerged. Seismic tomography has also led to an enhanced understanding of the core–mantle boundary region, which is of particular importance in SEDI studies. Together with new investigations on the nature of the thermal coupling between core and mantle, this boundary seems sure to be a focus for subsequent SEDI symposia.

Several new satellites devoted to mapping the geomagnetic field have recently been launched, and results are beginning to come onstream. If these projects prove to be as successful as the MAGSAT project, we can look forward to greatly improved geomagnetic data which can be used to infer dynamical behaviour in the core.

The last 6 years have also seen major developments in experiments motivated by the desire to understand processes going on in the Earth's core and mantle. The first working fluid dynamos, based on liquid sodium technology, have been constructed in the last few years. More such experiments are planned in the near future. Further progress has been made on high-pressure physics experiments, enhancing our knowledge of the properties of matter under core conditions. Less spectacular, perhaps, but nevertheless very valuable have been experiments on convection in rotating systems. These are developing our understanding of the dynamics of the core, a subject which up to now has only been amenable to theoretical treatment.

At the time of the 4th symposium, the first self-consistent convection-driven dynamo models were just beginning to come onstream. Since that time, this activity has expanded enormously, with many groups all over the world now involved. After this burst of activity, the last few years have seen a period of consolidation and interpretation, but few can doubt that numerical modelling of core and mantle convection will continue to play a major role in future SEDI activities. The study of reversals, for example, is now an activity which involves numerical modelling, paleomagnetic observations, and mantle convection studies. Mantle convection itself is another area where rapid progress is being made, in combination with results from seismic tomography, although the nature of convection in the lower mantle remains controversial.

Although many of the core and mantle processes evolve over very long time scales, the behaviour of short-period oscillations such as inertial modes can give useful information about the state of the core. They also give rise to some very interesting mathematical problems. Considerable progress has been made recently in this area by a combination of theory, numerical experiments and laboratory experiments.

The result of all this SEDI activity has been to develop greatly our understanding of the interior of planet Earth. A natural development will be to apply this knowledge to the interiors of other planets, though whether such studies can still be considered as SEDI activities will be for the organization itself to decide!

Acknowledgements

We are grateful to the University of Exeter, to the SEDI organization, and to the International Union of Geodesy and Geophysics (IUGG), for their support.

1 Thermal interaction of the core and mantle

Peter Olson

Department of Earth and Planetary Sciences, Johns Hopkins University, Baltimore, MD, USA

1. Introduction

The transfer of heat across the boundary between the core and mantle is fundamentally important, not just for the core and mantle individually, but also for the Earth as a whole. Core–mantle heat transfer determines the rate of growth of the solid inner core and, by controlling the rate at which energy is released to drive the geodynamo, exerts long-term control on the strength and the structure of the geomagnetic field. It also influences the structure and the dynamics of the heterogeneous D'' layer at the base of the mantle, and may have direct influence on surface processes including plate tectonics and continental breakup through generation of deep mantle plumes.

This chapter reviews recent progress in understanding the nature of core–mantle thermal interaction, with a focus on the long-term processes in the core–mantle boundary region. These processes occur on time scales where the effects of heat transfer are most significant. Other forms of core–mantle interaction, for which heat transfer is not of primary importance, typically occur on shorter time scales. Examples of shorter time scale processes include topographic coupling at the core–mantle boundary and gravitational coupling with the solid inner core. These are treated by D. Jault in Chapter 3 of this volume.

2. Conceptual background

The idea that heat transfer controls long-term core–mantle interaction is not new. Many of the basic concepts of core–mantle interaction, including thermal coupling, were proposed several decades ago (see Hide, 1970; Vogt, 1975; Jones, 1977). These concepts have been refined over the intervening years, as new observational constraints have become available (Gubbins and Richards, 1986; Bloxham and Gubbins, 1987; Gubbins and Bloxham, 1987; Bloxham and Jackson, 1990; Gubbins, 1997). But even with refinements, thermal coupling of the core and mantle remains just a hypothesis to this day, although an increasingly compelling one. Its appeal is based partly on the fact that it is founded on some very simple theoretical considerations, and partly because the weight of the observational evidence increasingly supports it.

The theoretical foundations underlying the core–mantle thermal coupling hypothesis are as follows. First, we know that both the mantle and core convect, and that thermal and compositional convection play a role in each. These two convective systems interact at the core–mantle boundary (here abbreviated CMB) where heat transfer occurs mostly by conduction, and mass transfer is suppressed, if not eliminated. Second, we know that the mantle and outer core convection have extreme differences in physical properties, as illustrated by the comparison of dimensionless parameter values in Table 1. The fluid outer core is an iron-rich

Table 1 Values of dimensionless parameters for the mantle and outer core

Parameter	Definition	Mantle	Outer core
Radius ratio		$r_c/r_o = 0.54$	$r_i/r_c = 0.34$
Prandtl number	$Pr = \nu/\kappa$	10^{23}	0.01
Ekman number	$E = \nu/\Omega D^2$	$\sim 10^9$	$\sim 10^{-13}$
Rayleigh number	$Ra = \alpha g \Delta T D^3/\kappa \nu$	$\sim 2 \times 10^7$	$\sim 10^{20}$
Supercriticality	Ra/Ra_c	$\sim 10^4$	$\sim 10^7$
Temperature heterogeneity	$S = \Delta_H T/\Delta_r T$	~ 0.2	$\sim 10^{-6}$
Reynolds number	$Re = UD/\nu$	10^{-20}	10^7
Peclet number	$Pe = UD/\kappa$	10^3	10^6
Nusselt number	$Nu = qD/k\Delta T$	20–30	$\sim 10^3$
Elsasser number	$\Lambda = \sigma B^2/\rho \Omega$	$\ll 1$	~ 1

alloy (Jeanloz, 1990; Stixrude *et al.*, 1997; Laio *et al.*, 2000). Even at the high pressures of the Earth's deep interior, the outer core fluid is nearly inviscid. In contrast, the flow in the subsolidus lower mantle is entirely dominated by its viscosity. Convection in the core is controlled primarily the Coriolis accleration, by Lorentz forces derived from the geomagnetic field, and secondarily by the intertia of the fluid, all of which are negligible in mantle convection. Mantle convection is mostly thermal, with compositional and phase change effects playing secondary roles (Tackley, 2000; Schubert *et al.*, 2001). In the outer core, compositional buoyancy produced by differentiation and solidification at the inner core boundary (abbreviated ICB here) probably dominates the convection, particularly with increasing depth (Buffett, 2000). Thermal convection in the outer core is closely linked to compositional convection, and in the critical region near the CMB, thermal convection may be the dominant form of motion. The time scales of convection are also vastly different in the two regions. The overturn time of mantle convection is measured in hundred million years; the overtime of convection in the core is measured in centuries.

The differences in force balances and time scales between the two systems mean that the thermal interaction between the core and the mantle is very unequal. As shown in the next section, the lower mantle responds to the CMB as an isothermal, free-slip surface, with negligible lateral temperature variations and negligible shear stress, whereas the outer core responds to the CMB as a rigid surface on which a pattern of heat flow is imposed by the slowly evolving dynamics in the overlying mantle. In this interaction, the core is more sensitive to the mantle than the mantle is to the core, because the core responds to both the average heat flow on the CMB as well as deviations from the average, whereas the mantle responds only to the average temperature there. These general concepts are agreed on. What is not agreed on is *how* the core and mantle respond to these conditions.

It is easy to construct a general theoretical argument for core–mantle thermal interaction on long time scales. Suppose, for example, that core–mantle interaction was entirely homogeneous, with heat transfer the same everywhere over the CMB. Then over long times, the azimuthal flow in the core would tend to average out any longitudinal differences, and the behavior of the geodynamo and the structure of the geomagnetic field would bear no relationship to the three-dimensional structure of the lower mantle. However, the observational evidence accumulated in the past decade indicates this is not the case. On the longest time scales, the existence of magnetic superchrons and the nonuniformity of transitional virtual geomagnetic poles indicates the geodynamo has a memory far longer than the time scales intrinsic to convection in the outer core. On somewhat shorter time scales, time-averaged geomagnetic field models indicate that some of the nonaxisymmetric structures seen in the historical field have persisted near their present-day locations for 5 Ma and perhaps longer.

To summarize, convection in the fluid outer core is linked to the structure of the lower mantle in several ways. The average thermal gradient at the base of the mantle determines the total heat flow from the core to the mantle, which in turn governs the cooling rate of the whole core, the rate of inner core solidification, and the power available from convection to drive the geodynamo. Convection in the core is also sensitive to lateral variations in lower mantle structure, especially to structure that produces lateral variations in heat flow at the CMB. Variations in radial heat flow at the CMB, resulting from lateral temperature gradients in the lower mantle, produce variations in the radial temperature gradient within the core. This interaction can alter the pattern of convection in the core, enhancing convection where the radial temperature gradients are large and suppressing convection where they are reduced. Because core convection is a major energy source for the geodynamo, modulation of the convection pattern should affect the structure of the geomagnetic field. According to this reasoning, there should exist some relationship between the pattern of mantle heterogeneity as revealed by seismic tomography, the flow in the outer core, and the structure of the geomagnetic field. Also, because the pattern of mantle heterogeneity evolves slowly, the coupling with the geomagnetic field should persist for millions of years.

The contents of this chapter are as follows. First, I describe the thermal environment of the CMB region as currently envisioned, based on simple heat transfer considerations. Second, I review the seismic structure of the lowermost mantle, and derive a pattern of heat flow variations on the CMB consistent with this structure. I then review experimental and numerical models of convection and dynamo action in the core subject to CMB thermal heterogeneity. Finally, I review the evidence for core–mantle thermal interaction from the geomagnetic and paleomagnetic field, and suggest some avenues for future progress.

3. Thermal boundary conditions at the CMB

Because the convective velocities in the liquid outer core are so much larger than in the solid mantle, the two regions respond very differently to the continuity of heat flow and temperature at the CMB. This situation is illustrated schematically in Figure 1. In physical terms, it is analogous to conditions on either side of the ocean floor, where the ocean crust and lithosphere are sensitive to the average temperature of the deep ocean, while the deep ocean is most sensitive to heat flow variations imposed on it at the sea floor. For thermal interaction near the CMB, the outer core plays a role similar to the ocean and the D″ layer of the mantle plays a role similar to the crust–lithosphere system.

The difference in response of the mantle and core in their thermal interaction can be demonstrated by considering the continuity of heat transport and temperature in the neighborhood of the CMB. In both the core and the mantle, it is convenient to represent the temperature as a sum of radial variations along an adiabat $T^{\mathrm{a}}(r)$ and departures from an adiabat Θ. The departure from adiabatic temperature consists of two parts, a spherical average $\overline{\Theta}(r)$ plus deviations from this average, Θ'. Wherever they appear together in this chapter, mantle and core quantities will be distinguished with subscripts, m for the mantle and c for the core. With this notation, the spherically averaged balance of heat flux near the CMB can be written as

$$\rho_{\mathrm{m}} C_{\mathrm{m}} \overline{w_{\mathrm{m}} \Theta'_{\mathrm{m}}} - k_{\mathrm{m}} \frac{\partial (T_{\mathrm{m}}^{\mathrm{a}} + \overline{\Theta}_{\mathrm{m}})}{\partial r} = \rho_{\mathrm{c}} C_{\mathrm{c}} \overline{w_{\mathrm{c}} \Theta'_{\mathrm{c}}} - k_{\mathrm{c}} \frac{\partial (T_{\mathrm{c}}^{\mathrm{a}} + \overline{\Theta}_{\mathrm{c}})}{\partial r}, \qquad (1)$$

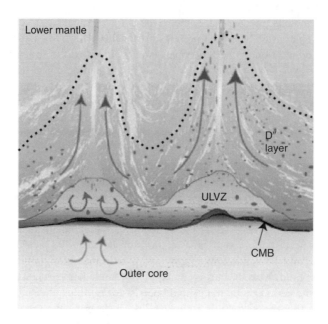

Figure 1 Schematic depiction of small-scale convection and sedimentation processes near the CMB proposed to explain the ULVZ. Possible mechanisms include iron silicate sediments filling inverted basins on the core side of the CMB, infiltration of core material into the mantle D″ layer, and partial melting of mantle material above the CMB. (Courtesy, E. Garnero.)

where ρ, C, and w denote density, specific heat, and radial velocity, respectively. An estimate for the relative magnitude of lateral temperature variations Θ' on either side of the CMB can be made by equating the two transport terms in Eq. (1). This gives

$$\frac{\Theta'_c}{\Theta'_m} \simeq \frac{\rho_m C_m}{\rho_c C_c} \frac{w_m}{w_c}. \tag{2}$$

Since $\rho_m C_m \simeq \rho_c C_c$ across the CMB, the ratio of lateral temperature variations is roughly the inverse of the ratio of radial fluid velocities on either side of the boundary. Typical estimates for these velocities are $w_m \sim 1$ cm/year (Tackley *et al.*, 1994; Schubert *et al.*, 2001) and $w_c \sim 0.1$ mm/s (Whaler, 1986; Bloxham and Jackson, 1991), which gives $\Theta'_c/\Theta'_m \sim 10^{-6}$. Thus, the lateral temperature variations in the core are negligible compared to those in the mantle. Insofar as its own dynamics, the mantle does not respond to the lateral temperature variations in the core. The mantle responds only to the spherically averaged temperature at the CMB. Accordingly, the continuity of temperature at the CMB $r = r_c = 3,480$ km:

$$T_m^a + \Theta_m = T_c^a + \Theta_c \tag{3}$$

reduces to an isothermal boundary condition *for the mantle*:

$$T_m = T_c^a + \overline{\Theta}_c, \quad r = r_c. \tag{4}$$

Assuming that heat is transported across the CMB by conduction only, the continuity of heat flow then reduces to a pair of heat flow boundary conditions *for the core*:

$$q_c = \overline{q_c} + q'_c, \quad r = r_c, \tag{5}$$

where

$$\overline{q_c} = -k_m \frac{d(T_m^a + \overline{\Theta_m})}{dr} \tag{6}$$

and

$$q_c' = -k_m \frac{\partial \Theta_m'}{\partial r}. \tag{7}$$

Condition (6) applies to the spherical average heat flow at the CMB, and condition (7) to lateral variations of heat flow. The equivalent expression for the spherical average CMB heat flow in terms of gradients in core temperature is just

$$\overline{q_c} = -k_c \frac{d(T_c^a + \overline{\Theta_c})}{dr}, \quad r = r_c. \tag{8}$$

Thus, the core and mantle are affected differently by conditions at the CMB. The mantle responds to the CMB as a free-slip isothermal boundary with temperature given by Eq. (4). The core responds to the CMB as a no-slip boundary with laterally variable heat flow given by Eqs (5)–(7).

Convection in the core is almost certainly turbulent. This assertion is made with confidence for two reasons. First, experiments show that rotating magnetoconvection in liquid metals becomes turbulent very near the critical Rayleigh number (Aurnou and Olson, 2001). Second, the Reynolds number Re for convection in the core is large, even for rather small scales of motion. For example, if we assume $U \sim 10^{-3}$ m/s, $\nu \sim 10^{-6}$ m^2/s and take $D \sim 10$ km for an eddy size, then $Re \sim 10^7$, as shown in Table 1. Turbulence implies the core will be well mixed where ever convection occurs, and the temperature gradient must very nearly match the adiabatic gradient. In convecting regions of the core, away from thermal boundary layers, the deviation of the mean temperature from the adiabat Θ_c' is comparable to the lateral variations of temperature,

$$\Theta_c' \sim q_c'/\rho C_c w_c, \tag{9}$$

where q_c' is the laterally varying part of the core heat flux, a reasonable estimate of the convective heat flux. Taking $q_c' \sim 2$ mW/m^2 and $w_c = 10^{-4}$ m/s in Eq. (9) gives $\Theta_c' \sim 10^{-4}$ K, a very small temperature variation, as shown previously by Stevenson (1987).

Because the heat conducted down the core adiabat represents a substantial fraction of the heat leaving the core, it is possible that the convective heat flux in the core is rather small compared to the total heat flow, and perhaps even small compared to the conductive heat flow. Some of the consequences of an approximate equality between heat conduction down the core adiabat and heat flow at the CMB have been investigated previously (Loper, 1978a,b; Lister and Buffett, 1995, 1998). One consequence is that rather unusual thermal regime can prevail near the CMB, which magnifies the role of lateral heat flow variations imposed by the mantle on the core.

The following is an example of this thermal regime. The adiabatic temperature gradient in the core near the CMB is given by

$$\frac{dT_c^a}{dr} = -\frac{\overline{T_c}}{H_T} \tag{10}$$

with $H_T = C_c/\alpha g_c$. Taking $\overline{T_c} = 4,000$ K for the core temperature near the CMB and $H_T = 5 \times 10^6$ m (Stixrude *et al.*, 1997; Boehler, 2000) gives $dT_c^a/dr \simeq -0.8$ K/km at the

CMB. Assuming the core conductivity $k_c \simeq 25\,\text{W/mK}$, the contribution to the heat flux at the CMB from conduction down the adiabatic temperature gradient is about $q_c^a = 20\,\text{mW/m}^2$. Now suppose that the total heat loss at the CMB is $3 \times 10^{12}\,\text{W}$, as assumed by Buffett *et al.* (1996) in their model of core thermal evolution. This is equivalent to an average heat flow on the CMB of $\overline{q}_c = 20\,\text{mW/m}^2$. In this example, the average convective heat flow in the outer core just below the CMB:

$$\overline{\delta q}_c = \overline{q}_c - q_c^a = 0. \tag{11}$$

It is possible that the net convective heat flow in the outer core near the CMB is actually negative. For example, if the total heat loss at the CMB is substantially less than $3 \times 10^{12}\,\text{W}$, then the average convective heat transfer near the CMB would be negative (downward), as discussed by Loper (1978a,b).

   If the total convective heat flux in the core is a small fraction of the total heat flow at the CMB, and if the heat flux variations are over the CMB are large, a consequence of mantle convection, then the lateral variations in the convective heat flow in the core may also be very large. Under these conditions, the CMB would consist of two types of regions, each covering about one half of the surface area. Beneath regions with higher than average CMB heat flow, the outer core fluid would be unstable to thermal convection and the convective heat flow in the core would be large and positive. Beneath regions with lower than average CMB heat flow, the outer core would be stable to thermal convection and the convective heat flow in the core would be large but negative. Beneath regions with average CMB heat flow, the outer core would be marginally unstable to thermal convection, and the convective heat flow would be small. In other words, the effect on the core of lateral heterogeneity in the D″ layer at the base of the mantle is to produce very large lateral variations in convective heat flow in the core.

4. CMB heat flow patterns from lower mantle seismic tomography

The analysis in the previous section suggests the possibility that heat flow variations on the CMB associated with convection in the lower mantle can result in very large variations in the convective heat flux in the core. How can this possibility be tested? One way is by estimating the heat flow variation on the CMB that is consistent with the three-dimensional seismic structure of the D″ layer in the lower mantle and known thermodynamic properties of lower mantle materials.

   The simplest approach to formulating heterogeneous heat flow boundary conditions at the CMB is to assume a linear correlation between heat flow variations and seismic velocity variations in the D″ layer (Olson and Glatzmaier, 1996; Glatzmaier *et al.*, 1999; Gibbons and Gubbins, 2000) of the form

$$q_c' = a\delta V_s / V_s, \tag{12}$$

where $\delta V_s / V_s$ is the shear wave heterogeneity at a particular depth and a is a coefficient of proportionality. The rationale for this relationship is the assumption that seismically fast regions in the D″ layer are cold downwellings, where the CMB heat flow is high, and seismically slow regions represent hot mantle, where the temperature gradient at the CMB and the heat flow are low. The coefficient a is difficult to constrain a priori. For example, its value depends on how much of the D″ layer seismic heterogeneity is assumed to be thermal, as opposed to compositional. In convection and dynamo models, the usual procedure is to choose a value for a arbitrarily.

A marginally better approach is to estimate the thickness of the D″ thermal boundary layer using standard boundary layer methods, and then calculate the CMB heat flow variations from the boundary layer thickness variations. The first step in this method is the same as in the one described above, conversion from seismic velocity variations to temperature variations. This requires an estimate of the correlation between seismic velocity variations and density variations from mineral physics, plus the assumption that all of the seismic heterogeneity in D″ is thermal in origin. The next step is to calculate the thickness of the assumed thermal boundary layer, using the seismic heterogeneity integrated through the D″ region. This technique gives the lateral variations of CMB heat flux q'_c relatve to an assumed average CMB heat flux $\bar{q}_c$.

One method to convert seismic velocity variations to temperature variations is to assume a constant value for the thermodynamic parameter β:

$$\beta = \frac{\partial (\ln V)}{\partial (\ln \rho)}, \tag{13}$$

where V is the seismic wave velocity (either V_p or V_s) and ρ is density in the D″ layer. It is convenient to introduce the integral scale for the thermal boundary layer thickness h of the D″ layer:

$$h = \int \frac{T - T_d}{\Delta T} dr, \tag{14}$$

where $\Delta T = T_c - T_d$ is the temperature difference across the D″ layer from the CMB radius r_c to r_d, the radius of the top of the D″ layer. In terms of h, the heat flow on the CMB is

$$q_c = q_m = k_m \Delta T / h \tag{15}$$

and the heat flow variations q'_m are related to the average heat flow $\bar{q}_m$ by

$$\frac{q'_m}{\bar{q}_m} = \frac{\bar{h}}{h} - 1, \tag{16}$$

where $\bar{h}$ represents the spherically averaged thermal boundary layer thickness of D″. Using Eq. (13), the average perturbation in seismic velocity through the D″ layer is then

$$\frac{\delta V}{V} = -\frac{\alpha \beta \Delta T}{\bar{h}} \int \frac{T - \bar{T}}{\Delta T} dr = \alpha \beta \Delta T \left(1 - \frac{h}{\bar{h}} \right), \tag{17}$$

where α is thermal expansivity. Combining Eqs (16) and (17) gives

$$\frac{q'_m}{\bar{q}_m} = \frac{\delta V/V}{\alpha \beta \Delta T + (\delta V/V)}. \tag{18}$$

Note that this relationship between heat flow variations and the depth-averaged D″ layer seismic wave velocity heterogeneity is not linear, although it is approximately linear when $\alpha \beta \Delta T \gg \delta V/V$.

Figure 2 shows the depth-averaged variations in shear wave velocity $\delta V_s / V_s$ in the D″ layer, obtained by averaging the shear wave velocity model SAW12d of Li and Romanowicz (1996) from $r_c = 3,480$ km to $r_d = 3,200$ km. This tomography model shows the prominent low velocity structures beneath the Southcentral Pacific and Africa that are often interpreted as high-temperature upwellings in the lower mantle (see Tackley, 2000 for a review of the

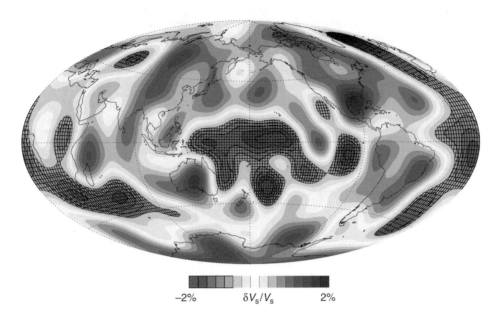

-2%　　　　　$\delta V_s / V_s$　　　　2%

Figure 2　Three-dimensional structure of the D″ layer at the base of the mantle. Contours of seismic shear wave velocity heterogeneity from seismic tomography model SAW12, averaged over the depth interval 2,700–2,880 km. The model SAW12 from Li and Romanowicz (1996) is truncated at spherical harmonic degree 12. Hatched areas indicate low velocity regions.

various dynamical interpretations of these). The model in Figure 2 also shows the prominent high-velocity ring surrounding the Pacific basin, often interpreted as cold mantle material, perhaps representing remnants of subducted lithospheric slabs. The peak-to-peak variation in depth-averaged shear wave heterogeneity in the D″ layer in this model is about 4%.

To convert the depth-averaged shear wave heterogeneity shown in Figure 2 to CMB heat flow variations, we must specify $\alpha\beta\Delta T$ and $\overline{q}_m$ in Eq. (18). For shear waves in the D″ layer, it has been estimated that $\beta = \partial(\ln V_s)/\partial(\ln \rho) \simeq 2.4$–3 and $\alpha\beta\Delta T \simeq 6 \times 10^{-2}$ although there is a large uncertainty in this parameter (Yuen *et al.*, 1993, 1996). The average heat flow from the core is equally uncertain, if not more so. A very conservative lower bound estimate comes from the total buoyancy flux at hotspot swells, and is based on the assumptions that hotspots result from mantle plumes originating in the D″ layer, and that plumes represent the only heat transfer from D″. This gives $4\pi r_c^2 \overline{q}_m \simeq 2.3$ TW for the total core heat loss (Schubert *et al.*, 2001), which less than the heat conducted down the core adiabat. In the other extreme, a geochemical model of the radioactive heat production in the whole Earth by Calderwood (2000) predicts $4\pi r_c^2 \overline{q}_m \simeq 21$ TW for the total core heat loss.

The impact of these different core heat loss estimates on the heterogeneity of CMB heat flow and its effect on convection in the core can be illustrated by comparing two different models, with total CMB heat flows well within the bounds discussed above. Figure 3 shows the pattern of heat flow on the CMB obtained using Eq. (18) with $\beta = 0.05$ and $4\pi r_c^2 \overline{q}_m = 3$ TW, a relatively low core heat loss estimate used to model the evolution of the core by Buffett *et al.* (1992, 1996). Figure 3(a) shows q_c and Figure 3(b) shows the residual heat flux $\delta q_c = q_c - q_c^a$, the difference between the heat flux at the CMB and the heat conducted down the core adiabat.

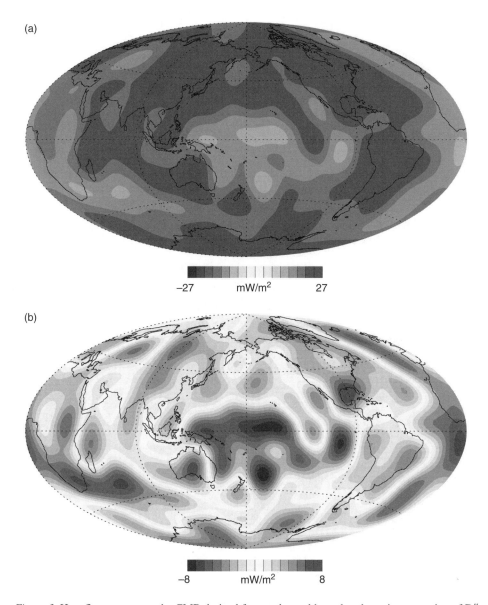

Figure 3 Heat flow pattern on the CMB derived from a thermal boundary layer interpretation of D''
layer seismic shear wave structure shown in Figure 2, assuming 3 TW total core–mantle
heat flow and a linear relationship between temperature and shear wave velocity variations.
(a) CMB heat flow pattern. (b) Residual CMB heat flow, the CMB heat flow minus the heat
flow conducted down the core adiabatic gradient. (See Colour Plate I.)

The actual variation in heat flow over the CMB is not very large in this model, and more
importantly, it is everywhere positive (i.e., heat flows from core to mantle). However, the
residual heat flux is positive over about half of the CMB and negative over the other half.
This residual CMB heat flux pattern is the one that convection in the core is sensitive to.

(a)

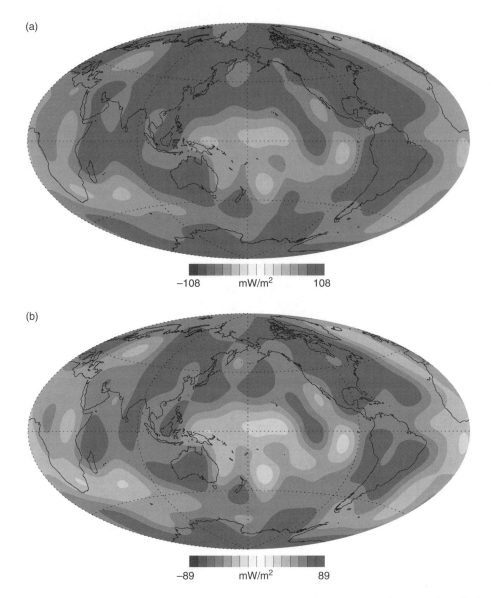

(b)

Figure 4 Heat flow pattern on the CMB derived from a thermal boundary layer interpretation of D″
layer seismic shear wave structure shown in Figure 2, assuming 12 TW total core–mantle
heat flow and a linear relationship between temperature and shear wave velocity variations.
(a) CMB heat flow pattern. (b) Residual CMB heat flow, the CMB heat flow minus the heat
flow conducted down the core adiabatic gradient. (See Colour Plate II.)

In the low heat flow regions beneath the Central Pacific and Africa, the residual heat flux
in the core near the CMB is negative. According to this model, these regions would not be
able to support free thermal convection in the outer core. Only where the residual heat flux
is positive would the CMB heat flow be able to support free thermal convection in the outer

core. These regions are mainly within longitude sectors beneath East Asia and Australia, and beneath the Americas.

Figure 4 shows the patterns of heat flow and residual heat flow on the CMB obtained using a high core heat loss estimate, $4\pi r_c^2 \bar{q}_m = 12\,\mathrm{TW}$. In this case, the heat flow and the residual heat flow are everywhere from core to mantle. According to this model, every region on the CMB would be able to support free thermal convection in the outer core. For this case, the CMB heat flow heterogeneity would only be expected to modulate the intensity of core convection. Further increases in total CMB heat flow, toward the 21 TW advocated by Calderwood (2000) would make the CMB appear even more homogeneous in heat flow.

In summary, these calculations indicate that the effect of heterogeneous CMB heat flow on convection in the outer core is expected to be strongest if the total CMB heat flow is comparable to or less than about 3 TW. Then, the local heat flow from the core to the mantle will be less than the heat conducted down the core adiabat over approximately one half of the CMB. These regions will be unable to support free thermal in the outer core. Only beneath the high heat flow regions on the CMB would free thermal convection occur in the core. In such a situation the convection in the core would be *heterogeneous*. Alternatively, if the total heat flow from the core is high, each region on the CMB can support free thermal convection in the outer core, and the convection in the outer core would be essentially homogeneous, but modulated by the D″ layer structure.

5. The thermochemical environment near the CMB

The standard, highly idealized view of the CMB is a sharp interface, with a nearly discontinuous change from the solid silicate lower mantle to the liquid iron-rich outer core. But recent discoveries suggest that the actual transition between the core and the mantle may in fact be far more complex. The cartoon in Figure 1 summarizes some of the smaller scale process that could affect the CMB region, and produce a more gradual transition.

Evidence for a transitional layer between the mantle and core is provided by the so-called ultra-low velocity zones, the ULVZs. These are thin patches of mantle with anomalously low seismic velocity (with seismic velocity reduced by nearly 10% relative to D″) located at or very near the CMB. Several of these patches have been identified. The best documented one is located beneath the Central Pacific (Garnero *et al.*, 1998) and more recently another has been identified beneath Southern Africa (Wysession *et al.*, 1999). The presumed relationship between ULVZ, the thicker D″ layer, and possible CMB chemical processes are shown in Figure 1. Originally the ULVZ were interpreted to be lenses of partially molten lower mantle material (Williams and Garnero, 1996). More recently, however, they have been given a new interpretation, based on the analogy between the CMB and the ocean floor (Buffett *et al.*, 2000). According to this interpretation, chemical precipitation of silica-rich, relatively low density solid particles occurs in the outer core. Buffett *et al.* (2000) argue that this precipitation is an expected consequence of inner core growth. As the iron-rich inner core solidifies, the iron concentration of the outer core fluid decreases, until it becomes supersaturated with respect to silicates. In this scenario, the low-density silicate sediments rise buoyantly through the outer core fluid, and are deposited in core–mantle basins, located at topographic highs on the CMB. Although little is known about the topography on CMB (Loper and Lay, 1995), it seems inevitable that some topography exists there, given the large seismic heterogeneity in the D″ layer. The lateral density variations implied by the seismic heterogeneity in the D″ layer would support topography variations on the CMB by the same mechanisms that thickness variations of the crust and temperature variations in upper mantle support topography at the

Earth's surface. And like surface topography, the link between heat flow and topography on the CMB would depend on the origin of the density variations in D″.

Core sedimentation has implications for the interpretation of heterogeneity throughout D″. For example, suppose the density heterogeneity in D″ is controlled by composition. By analogy with the surface elevation of the continental crust, the CMB will be depressed downward into the outer core where D″ is thick, forming an inverted highland. Heat flow from core to mantle will be low there. Sediments from the core would be shed from this highland and would collect in basins where D″ is thinner and the depression of the CMB is less. Heat flow from core to mantle is relatively high in such places. Thus, the inverted crust model of D″ predicts sedimentary ULVZs form in basins with relatively high CMB heat flow. Alternatively, suppose that D″ heterogeneity is controlled by temperature variations. By analogy with spreading ridges in the oceanic lithosphere, the CMB will be depressed downward into the outer core where the CMB heat flow is high. Sediments from the core would then be shed from these inverted highs and would tend to accumulate in distant core–mantle basins, where the heat flow is relatively low. Thus, the thermal boundary layer model of D″ predicts sedimentary ULVZs form in basins where the CMB heat flow is low.

According to the thermal boundary layer model of D″, sedimentary ULVZ should accumulate in basins where CMB heat flow is low. This association can be tested by comparing the location of known ULVZ with the CMB heat flow pattern shown in Figure 3, which was derived on the assumption that large-scale D″ seismic heterogeneity has a thermal origin. The best-delineated ULVZs are located beneath the Southcentral Pacific and beneath Southern Africa, places where the thermal boundary layer model of D″ predicts low CMB heat flow. Although this is by no means a definitive test, the results offer some support for a thermal interpretation of D″ seismic heterogeneity.

6. Theoretical and numerical models of thermal core–mantle interaction

Numerical and theoretical models of thermal convection and magnetoconvection in rotating spheres have shown that the influence of CMB thermal heterogeneity on the geodynamo may be quite significant. A review of published models on this topic also reveals a wide range of behaviors, an indication of the difficulty in predicting a priori how the geodynamo actually responds to heterogeneous mantle heating. To date, most of these studies have focused on finding conditions necessary for locking the nonaxisymmetric part of core convection and, in magnetoconvection and convection-driven dynamos, the nonaxisymmetric part of the magnetic field, to the mantle heterogeneity pattern. Locking is just one part of core–mantle thermal interaction, but even this seemingly straightforward issue has not yet to be fully resolved.

The early numerical models of rotating convection with inhomogeneous boundary heating by Zhang and Gubbins (1992, 1993, 1996) found that the azimuthal drift rate of the convection pattern is indeed affected by longitudinal thermal heterogeneity on the boundary. For some cases, particularly near the critical Rayleigh number, Zhang and Gubbins found the azimuthal drift rate vanishes and the convection pattern becomes locked to the boundary heterogeneity. Subsequent investigations have sought to determine whether the locking seen near the onset of motion persists to highly supercritical Rayleigh numbers, in both thermal convection (Sun et al., 1994) and magnetoconvection (Olson and Glatzmaier, 1996). Here the results were not so clear-cut. Parts of the flow in the outer portion of the shell were seen to lock to the boundary heterogeneity, but deeper within the shell the convect ion was relatively unaffected, and continued to drift azimuthally. Yoshida and Shudo (2000) made an analytical study of a linear response of the outer core flow to a thermally heterogeneous mantle. They considered the inviscid response of a rotating fluid half space to a Gaussian-shaped boundary temperature variation, and obtained a large-scale pattern similar to the experiment by Sumita and Olson (1999) shown in Figure 9, including an eastward flow near the thermal anomaly, and an eastward phase shift of the convection pattern relative to the nonrotating case. Nonlinear transport effects are expected to be important in convection in the core, which can lead to the formation of jet and front structures in the outer core, similar to those found in the ocean and the atmosphere. Localized jet structures are not prominent in the current generation of numerical dynamo models. But the appearance of these structures in numerical dynamo models is likely in the future, as their spatial and temporal resolution of the convection increases.

It is readily appreciated why it is so difficult to predict from first principles how the core responds to a boundary heat flow anomaly. If, for example, the anomalous heating is purely zonal, then the core response will be dominated by zonal thermal winds. Zonal winds tend to unlock the core and mantle. Alternatively, if the anomalous heating is predominantly sectorial, locking can occur, but even in this idealized case the nature of the locking depends on the strength of the variable boundary heating relative to other parameters of the convection.

Examples of fully locked thermal convection are given in Figures 5 and 6, showing the flow structure obtained by Gibbons and Gubbins (2000) from numerical models of rotating convection with variable boundary heat flux. Figure 5 shows the structure of steady convection driven by a spherical harmonic degree two heating pattern in an infinite Prandtl number fluid, for various Ekman numbers. Examination of the equatorial planforms shows a progressive eastward shift of the convection pattern relative to the boundary heating pattern as the rotation rate is increased. At low rotation rates, that is, larger E, the radial motion is negatively correlated with boundary heat flow. Downwellings occur beneath the high boundary heat flow sectors, where the core fluid temperature is lowest. As the rotation rate is increased, the flow structure becomes progressively more columnar, and in the equatorial plane, the planform shifts progressively further eastward. At the highest rotation rates considered ($E = 10^{-4}$), the downwellings are east of the high heat flow sectors, and close to the longitude where the boundary heat flow anomaly is zero. In this case, there is a strong eastward azimuthal flow below the high boundary heat flow sectors, and weaker westward flows below the low boundary heat flow sectors.

Another steady state convection pattern obtained by Gibbons and Gubbins (2000) is shown in Figure 6, using a more complex boundary heating pattern. In this case, the boundary heat flow pattern is linearly proportional to shear wave velocity heterogeneity in the D'' layer, as discussed in Section 4, the Ekman number is $E = 10^{-3}$ and the buoyancy number $B = E Ra = 1$. The steady flow pattern at the outer surface (the CMB is treated as a free-slip surface here) shows the same basic pattern as in the high rotation rate cases in Figure 5, namely an eastward phase shift of the convection pattern relative to the boundary heating by a quarter wavelength. Strong eastward equatorial flows are generated beneath the regions with highest heat flow. Also note the strong convergence of the azimuthal flow to the east of the high heat flow sectors, where the east and west azimuthal flows collide.

The presence of a magnetic field and strong Lorentz forces can alter this picture, and alter the longitudinal phase relationship between the convection pattern and the boundary heating. For example, Yoshida and Hamano (1993) showed, using a linear model of rotating magnetoconvection, that the eastward phase shift of the flow relative to the heating is suppressed when the Elsasser number $\Lambda \simeq 1$.

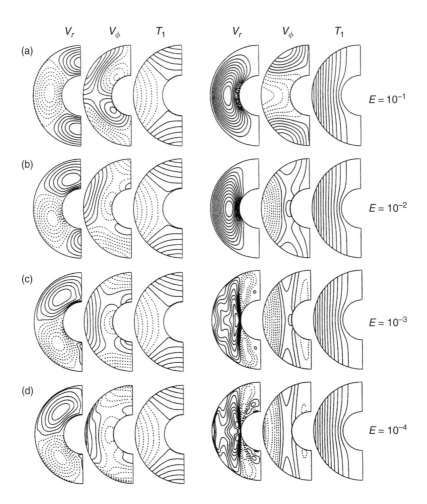

Figure 5 Steady state numerical solutions of thermal convection in a rotating spherical shell including heterogeneous heat flow at the outer boundary, at buoyancy number $B = RaE = 1$, for various Ekman numbers E in an infinite Prandtl number fluid. The variable boundary heating has a spherical harmonic $Y_2^2(\cos 2\phi)$ pattern, with amplitude $S \simeq 1$. Left three columns show equatorial plane sections (longitude increasing from top to bottom in each image) of radial velocity, azimuthal velocity, and temperature perturbations, respectively; right three columns show meridional plane sections at longitude $0°$ of the same three variables. Solid contours $=$ positive values; broken contours $=$ negative values. (Reprinted with permission from Gibbons, S. J. and Gubbins, D., "Convection in the Earth's core driven by lateral variation in the core–mantle boundary heat flux," *Geophys. J. Int.* **142**, 631–642, figures 1 and 9. Copyright (2000).)

The relationship between magnetic field, fluid velocity, and boundary heating patterns from a steady state numerical dynamo model by Sarson *et al.* (1997) is shown in Figure 7. In this example, the heating pattern varies as $\sin\theta \cos 2\phi$, and the azimuthal variation of all the variables is limited to wavenumbers 0 and 2, a so-called 2.5-dimensional numerical dynamo model. Here, the fluid upwelling and boundary heat flow are nearly anticorrelated. There is a slight westward phase shift of the downwellings relative to the high boundary

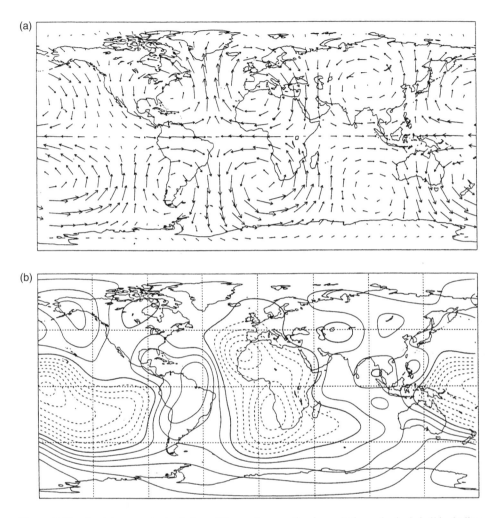

Figure 6 Steady state numerical solution of thermal convection in a rotating spherical shell including heterogeneous heat flow at the outer boundary, at buoyancy number $B = RaE = 1$, Ekman number $E = 10^{-3}$. The variable boundary heating pattern (b) is derived from lower mantle seismic heterogeneity, with assumed amplitude $S \simeq 1$. Heat flow highs have solid contours; heat flow lows have broken contours. The flow pattern is shown with velocity arrows in (a).

heat flow sectors. This correlation is reflected in the magnetic field intensity pattern on the outer spherical boundary, shown in Figure 7. The high-intensity flux lobes are located over the fluid downwellings, close to but slightly westward of the boundary heat flow maxima. Similarly, the longitude bands with low magnetic field intensity are nearly coincident with low boundary heat flow sectors, where the radial fluid motion is upwelling, except for the small westward phase shift.

Taken together, the results in Figures 5–7 indicate how steady state core flow can be locked to CMB heat flow heterogeneity. The all-important phase relationship between the boundary heat flow pattern and the flow depends, according to these results, on the relative strength of the Lorentz and Coriolis forces, that is, on the Elsasser number Λ. For small Λ, rotation

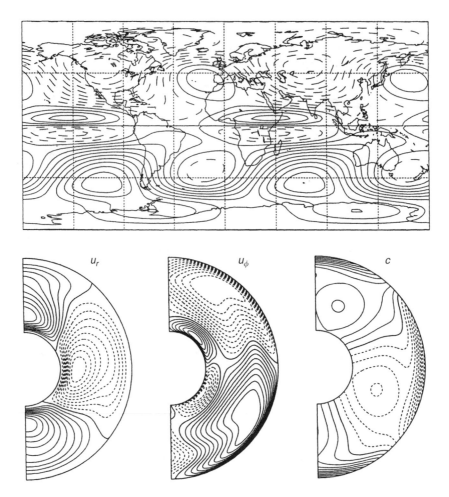

Figure 7 Coupling between boundary heat flow, core flow, and magnetic field in a steady state numerical dynamo model with heterogeneous heat flow boundary conditions. The boundary heat
flow has an azimuthal wavenumber 2 pattern with maxima at longitudes 0, 180°, and amplitude $q'/\bar{q} = 2$. Map: contours of radial magnetic field on the CMB. Bottom: equatorial
section contours of radial velocity, azimuthal velocity, and co-density (buoyancy), from longitude 0° (top) to 180° (bottom). In this dynamo model, core upwellings correlate with high
heat flow and low magnetic field intensity. The magnetic field intensity pattern is shifted
~35° westward relative to the boundary heating pattern. (Reprinted from *Phys. Earth Planet.
Inter.* **101**, Sarson, G. R., Jones, C. A., and Longbottom, A. W., "The influence of boundary
region heterogeneities on the glodynamo," 13–32, figures 10 and 11. Copyright (1997) with
permission from Elsevier Science.)

shifts the convection pattern to the east, whereas for larger Λ the Lorentz forces tend to
reduce or even slightly reverse the eastward phase shift. Another important result, which
does not show a phase shift, is the anticorrelation between the azimuthal variations of the
magnetic field and the flow pattern. High-intensity field on the boundary correlates with fluid
downwellings, and low-intensity field on the boundary correlates with fluid upwellings. This
particular relationship will be examined further later in this chapter.

A major limitation of the calculations just described is the steady state assumption. At lower Ekman numbers and higher Rayleigh numbers, the convection becomes more time dependent, and strictly steady states disappear. The response intrinsically time-dependent numerical dynamo models to variable boundary heating is, therefore, more difficult to characterize than steady state models. One example of the response of a highly time-dependent numerical dynamo to boundary heat flow heterogeneity is shown in Figure 8. This shows the pattern of

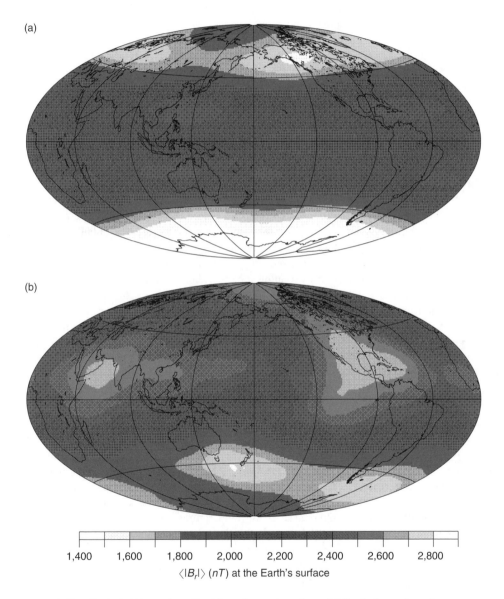

Figure 8 Nondipole field intensity at Earth's surface averaged over 20 dipole decay times from numerical dynamo models shown in Figure 16. (a) Dynamo model g with uniform heat flow at the outer boundary. (b) Dynamo model h with mantle tomography pattern of heat flow at the outer boundary. (See Colour Plate III.)

nondipole field intensity variations at Earth's surface averaged over a time period equivalent to 20 dipole decay times, from two numerical dynamo models by Glatzmaier *et al.* (1999) shown in Figure 15. Figure 8(a) is from Glatzmaier *et al.* dynamo model g with uniform heat flow at the outer boundary; Figure 8(b) is from Glatzmaier *et al.* dynamo model h, which uses a pattern of heat flow at the outer boundary proportional to lower mantle seismic heterogeneity. With uniform boundary heat flow (Figure 8(a)), the time average nondipole field intensity distribution is nearly zonal, as expected from symmetry considerations. However, with mantle tomographic boundary heat flow (Figure 8(b)), the nondipole field is most intense where the heat flow is highest. This relationship indicates that, even with highly time-dependent dynamo action, the time-averaged nonaxisymmetric field is concentrated where the boundary heat flow is high. A qualitatively similar result in a time-dependent numerical dynamo model has also been found by Bloxham (2000). The concentration of magnetic flux at these locations occurs either residual fluid downwellings, by intensified convection, or by both mechanisms acting together.

7. Experiments on thermal core–mantle interaction

There have been several experiments designed to model rotating spherical convection with heterogeneous temperature or heat flow boundaries (Hart *et al.*, 1986; Bolton and Sayler, 1991; Sumita and Olson, 1999). Most experiments of this type use the Busse and Carrigan (1976) technique of rapid rotation, which provides both a large Coriolis acceleration and simultaneously, through the centrifugal effect, buoyancy forces directed radially outward. A modification of this experiment is to use only the southern hemisphere and a rotation rate that results in an approximately radial buoyancy force orientation, as shown in Figure 9. Laboratory experiments using rapidly rotating spherical shells with low-viscosity fluids offer several advantages, as well as several disadvantages, in comparison with numerical models. It is possible to reach more realistic combinations of Rayleigh and Ekman numbers with experiments than with calculations, and with experiments there are none of the problems associated with finite resolution in numerical models. Thermal convection in these experiments typically contain many fine-scale, nearly two-dimensional columnar plumes, a form of quasi-geostrophic turbulence, in addition to larger scale motions such as azimuthal zonal flows.

Figure 9 shows a comparison of convection planforms with uniform and nonuniform heating on the outer boundary, from experiments by Sumita and Olson (1999). In these experiments, $E = 4.7 \times 10^{-6}$ and the effective gravity vector is directed nearly radially outward over much of the outer boundary, in order to minimize thermal wind effects originating from the basic state radial temperature gradient. A temperature difference is maintained between the inner and outer hemispherical boundaries and in addition, a strip heater is placed on the outer boundary at the position indicated in the figures. In order to simulate the correct buoyancy in a self-gravitating spherical shell, the temperatures are reversed in the experiment. The inner boundary, modeling the ICB, is cooled and the outer boundary, modeling the CMB, is heated. The heater strip on the outer boundary then represents anomalously cold mantle material with locally high CMB heat flux. The dye image in Figure 9(a) shows the structure of fine-scale quasi-geostrophic convection found at low Ekman number and highly supercritical Rayleigh number (here $Ra/Ra_c \simeq 19$) with uniform temperature on the outer boundary. There is a weak, westward azimuthal flow superimposed on the convection in this case.

Figure 9(b) and (c) shows how the convection planform is altered by anomalous boundary heating. Figure 9(b) shows the convection pattern at $Ra/Ra_c \simeq 25$, from an experiment

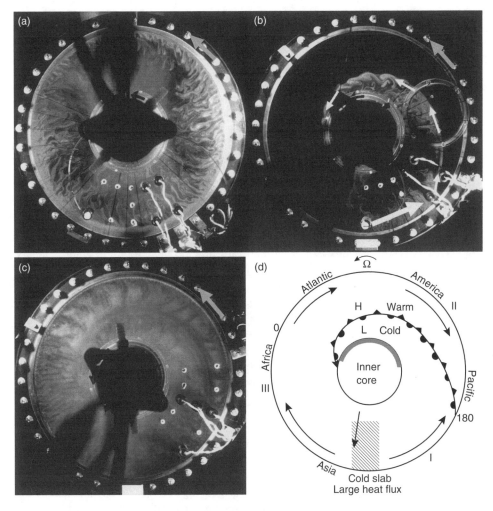

Figure 9 Convective patterns from experiments in a rotating hemispherical fluid shell including heterogeneous heating on the outer boundary, at Ekman number 4.7×10^{-6} and Prandtl number $Pr = 7$. Images are equatorial sections of columnar motion, viewed along the rotation axis. Rotation direction is counterclockwise, as indicated by the gray arrows on the equatorial rim. The location of the anomalously high boundary heat flow heterogeneity is shown by the white rectangle along the lower rim of each image. The convection planform is visualized by flakes (c) and fluorescent dye (a, b). Image (a) shows small-scale columnar thermal convection at supercritical conditions $Ra/Ra_c = 19$ with uniform thermal boundary conditions (no anomalous boundary heat flow). Images (b) and (c) show the convection planform the globally locked regime at $Ra/Ra_c = 26$ and 25, respectively, with strongly heterogeneous boundary heat flow. White arrows indicate the eastward azimuthal flow from the high heat flow region into the inward spiraling jet. (d) Schematic diagram of the flow pattern in the global locking regime applied to Earth's core. The shaded area on the ICB is the inferred high heat flow region produced by the inward spiraling jet. Warm and cold fluid is indicated by black and white arrow heads, respectively. The temperature front and jet are indicated by triangles and semi circles. (Reprinted with permission from *Science* **286**, Sumita, I. and Olson, P., "A laboratory model for convection in the Earth's core driven by a thermally heterogeneous mantle," 1547–1549, figures 2 and 3. Copyright (1999) American Association for the Advancement of Science.)

where the ratio of total heat flow at the outer boundary to the inner boundary is $Q^* = 1.57$. Figure 9(c) shows the convection pattern at $Ra/Ra_c \simeq 26$ and $Q^* = 2.17$. The primary change in the flow structure with heterogeneous boundary heating is the appearance of large, spiral-shaped cyclonic and anticyclonic eddies that are fixed with respect to the high heat flux patch. Note the general eastward phase shift of the flow pattern relative to the heated boundary region. Small-scale, time-dependent convection is superimposed on this larger scale stationary flow pattern. In addition to the large scale flows, an intense, concentrated jet spirals inward from the CMB toward the ICB to the east of the high heat flux region. This jet is clearly seen in images (b) and (c) in Figure 9 and separates counter-rotating warm and cold fluid eddies on either side. The jet serves to transport heat directly from the CMB to the ICB in the model. In this regard, it plays a dynamical role similar to western intensified currents in the general circulation of the ocean, such as the Gulf Stream. The dye pattern in Figure 9(b) shows that the jet begins as a laminar current and develops baroclinic shear instabilities as it approaches the ICB, reminiscent of the Gulf Stream instabilities. As opposed to the convection with homogeneous heat flux boundaries, where the zonal azimuthal flow is always westward, cases 9(b)- and 9(c)-driven heterogeneous CMB heat flux exhibit an eastward flow originating at the high boundary heat flux region, and westward flow where the heat flux is lower. These directions are qualitatively consistent with the results from numerical models with strong rotation and strongly heterogeneous boundary heating. The alternating directions of the azimuthal flow results in a convergence zone located east of the heated region, and it is here that the spiral jet forms. It also leads to an azimuthal divergence westward of the heated boundary region.

Based upon the results of these experiments, Sumita and Olson (1999) suggested an interpretation of flow in the outer core driven by D″ layer heat flow anomalies, shown schematically in Figure 9(d). The region on the CMB with the highest anomalous heat flow on the CMB is likely to be beneath East Asia, where, according to many seismic tomography models (Li and Romanowicz, 1996; van der Hilst and Kárason, 1999; Kuo *et al.*, 2000), the single largest patch with significantly higher than average seismic velocities seen in Figure 2. Interpreting the response of the core to this boundary heterogeneity in terms of the flow pattern shown in Figure 9, we expect there are basically two types of azimuthal flow beneath the CMB: a low-pressure region with a cold eastward flow originating beneath East Asia driven by the high heat flux from the mantle there, and a high-pressure surrounding region with a westward mean flow. Azimuthal convergence is predicted to occur eastward of the East Asia seismic anomaly, and azimuthal divergence to its west. This basic pattern of azimuthal flow is similar to the flow pattern inferred from geomagnetic secular variation using the tangentially geostrophic approximation (Bloxham and Jackson, 1991).

8. Nonaxisymmetric geomagnetic field structure on the CMB

Certain features in the nonaxisymmetric part of the geomagnetic and paleomagnetic fields have been cited as examples of thermal core–mantle coupling. One example is the pattern of magnetic field intensity at high latitudes. Figure 10 shows the time average historical field model by Bloxham and Jackson (1992) from 1870 to 1990, truncated at spherical harmonic degree 4. As many authors have pointed out, much of the geomagnetic field on the core–mantle boundary during historical times is concentrated into three or four high intensity patches. These high-latitude patches of concentrated flux occur in pairs, located

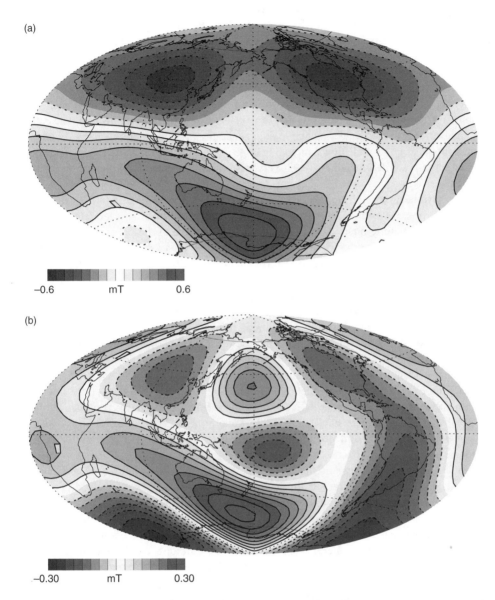

Figure 10 Time average geomagnetic field model on the CMB for 1870–1990 by Bloxham and
Jackson (1992). (a) Radial magnetic field intensity truncated at spherical harmonic degree
$l = 4$, the solid contor curve corresponds to the magnetic field equator. (b) Nonaxisymmetric radial magnetic field intensity. Solid and dash contours indicate positive and negative
values, respectively.

approximately at the same longitude in each hemisphere. It has been proposed that these
high intensities are produced by quasi-geostrophic, columnar-style convection in the core
(Gubbins and Bloxham, 1987). The locations of the flux patches, and by inference the location of the flow structures that induce them, have persisted for 250 years or more (Bloxham

and Jackson, 1992). Bloxham and Gubbins (1987) proposed that persistent, nearly stationarity flux lobes are the result of thermal core–mantle interaction. Numerical dynamo models explain these patches as concentrations of smaller scale flux patches, produced by low Ekman number columnar convection in the region of the core outside the inner core tangent cylinder (Christensen *et al.*, 1999). Stationary, high-intensity magnetic flux patches in the core could be produced by two mechanisms, both related to CMB thermal heterogeneity. First, the convection is expected to be more intense beneath high CMB heat flux regions, for the reasons discussed in the previous sections. Locally intensified convection should result in locally stronger magnetic field generation. Second, heterogeneous boundary heating produces large-scale flows, including large-scale downwellings. As shown in the next section, large-scale downwellings, even very weak ones, can readily explain these intense flux patches.

There have been several models of time-averaged magnetic field structure on the CMB that contain nonaxisymmetric field components which could also be interpreted in terms of thermal core–mantle interaction, and which are suggestive of the fields produced by the heterogeneously heated numerical dynamo models. Figures 11–13 show time-averaged paleomagnetic field models by Johnson and Constable (1995, 1997), Kelly and Gubbins (1997), and Kono *et al.* (2000b), respectively. The first two use globally distributed paleomagnetic directions recorded in lavas and sediments to represent the time-averaged paleomagnetic field for the past 5 Ma, and are truncated at spherical harmonic degree 4. The model of Kono *et al.* (2000b) uses intensity and direction data, and is truncated at spherical harmonic degree 3. The normal polarity fields at the CMB from these models are shown in the figures.

There are some significant differences in these paleomagnetic field models, which are particularly evident in the nonzonal field patterns. Like the historical field model of Bloxham and Jackson (1992), the paleomagnetic field model by Kelly and Gubbins (1997) includes prominent, dense flux patches at high latitudes in both hemispheres. In contrast, the field models by Johnson and Constable (1997) and Kono *et al.* (2000b) have relatively weaker nonzonal fields overall, and do not show the same flux bundle concentrations at high latitudes.

Genuine differences between the historical and the paleomagnetic field models are expected, given the great difference in the length of the time averages involved. The time span of the historical field is far shorter than the characteristic 15,000-year dipole decay time, whereas the paleofield models span many dipole decay times. Even so, the three paleofield models agree with each other and with the historical field model in some important respects. One place the models show general agreement is in the longitude sector beneath the Australia and Southeast Asia. In all the models the contours of the total radial magnetic field diverge beneath the Central Pacific, indicating relatively low nonaxisymmetric field intensity there. To the west of this low, the contours in all the field models converge toward the equator, indicating relatively high nonaxisymmetric field intensity there. Johnson and Constable (1998) have analyzed the directional data that define this anomalous structure, using paleomagnetic measurements for 0–5 Ma and archeomagnetic and paleomagnetic measurements for 0–3 ka. They conclude that the anomaly is genuine. A time-dependent archeomagnetic field model covering the last 3,000 years, derived from a larger set of directional measurements, also contains this pattern of nonaxisymmetric field (Constable *et al.*, 2000). It has been pointed out that the low-field region in the Central Pacific seen in Figures 10–13 coincides with the broad scale Pacific velocity low seen in many seismic tomographic images of the lower mantle, as in Figure 2, and with the ULVZ in the same region (Garnero *et al.*, 1998), and may be a product of core–mantle interaction.

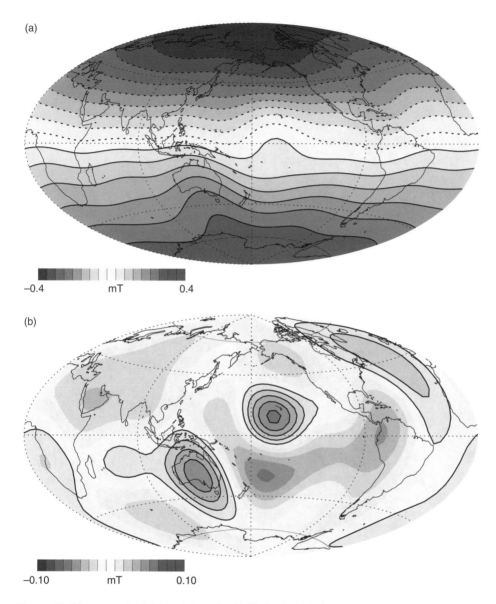

Figure 11 Paleomagnetic field model on the CMB for 0–5 Ma by Johnson and Constable (1995, 1997). (a) Radial magnetic field truncated at spherical harmonic degree 4, solid contour corresponds to the magnetic field equator. (b) Nonaxisymmetric radial magnetic field intensity. Solid and dash contours indicate positive and negative values, respectively.

The high field intensity region to the west of the Pacific, beneath Southeast Asia and Australia, is also a part of this structure but it has not received the same attention, even though it is the feature most common to all the time-averaged field models. It also may be a product of core–mantle interaction. This field intensity high lies beneath and slightly

(a)

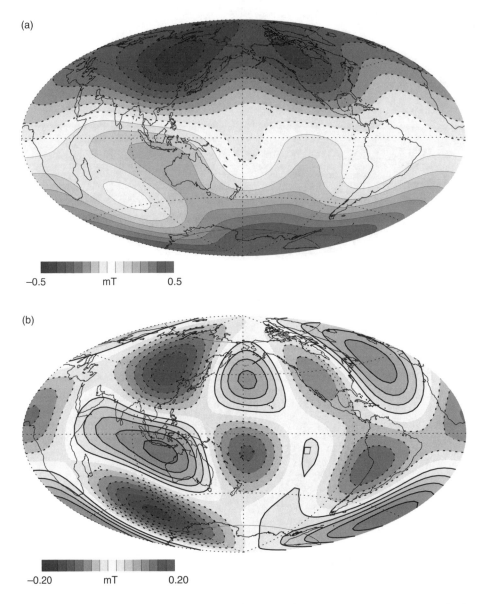

Figure 12 Paleomagnetic field model on the CMB for 0–5 Ma by Kelly and Gubbins (1997). (a) Radial
magnetic field truncated at spherical harmonic degree 4, solid contour corresponds to the
magnetic field equator. (b) Nonaxisymmetric radial magnetic field intensity. Solid and
dash contours indicate positive and negative values, respectively.

to the east of the high seismic velocity region in the D″ layer, where, as discussed previously in this chapter, high CMB heat flow is suspected. According to the numerical
dynamo models, high CMB heat flow is expected to result in anomalously high field
intensity in this sector, through residual fluid downwellings or by enhanced small-scale
convection.

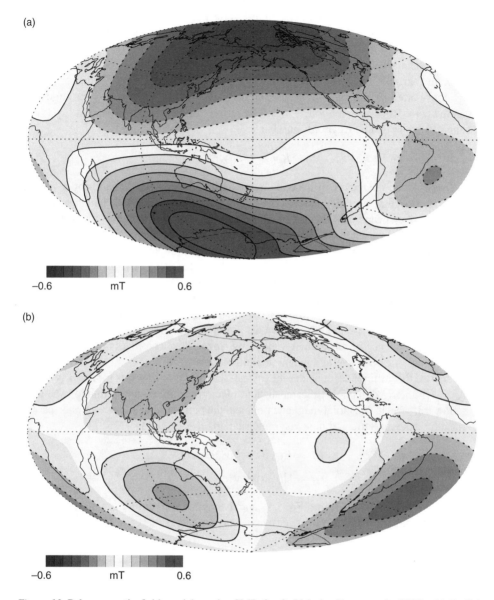

Figure 13 Paleomagnetic field model on the CMB for 0–5 Ma by Kono *et al.* (2000). (a) Radial
magnetic field truncated at spherical harmonic degree 3, solid contour corresponds to the
magnetic field equator. (b) Nonaxisymmetric radial magnetic field intensity. Solid and
dash contours indicate positive and negative values, respectively.

9. Magnetic images of core flow

Methods for imaging core flow use some approximation to the radial induction equation
(Roberts and Gubbins, 1987)

$$\frac{\partial B_r}{\partial t} + (\mathbf{U}_H \cdot \mathbf{\nabla}_H)B_r + B_r(\mathbf{\nabla}_H \cdot \mathbf{U}_H) = \frac{\lambda}{r}\nabla^2(r B_r) \qquad (19)$$

to infer the fluid velocity tangent to the CMB $\mathbf{U}_H$ in terms of the radial component of the magnetic field B_r and its time derivative. Here λ is the magnetic diffusivity of the outer core fluid, assumed constant, t is time, r is the radial coordinate and the subscript H indicates the tangential spherical coordinates (θ, ϕ).

The traditional approach, the so-called frozen flux method, neglects the diffusion term in Eq. (19), and assumes the outer core fluid acts like a perfectly electrical conductor. Simple scaling considerations indicate that neglecting magnetic diffusion is justified for large scale flow in the core (Roberts and Scott, 1965). The ratio of transport to diffusion in Eq. (19) is measured by the magnetic Reynolds number

$$Rm = UD/\lambda, \tag{20}$$

where U and D are characteristic velocity and length scales, respectively. Using a velocity estimated from secular variation and the core radius r_c for the characteristic length scale gives $Rm \simeq 1{,}000$. However, since this approach relies on the secular variation of the magnetic field, it is not useful for interpreting long-term average field models.

An alternative approach is to assume a mean field electrodynamical balance. Mean field electrodynamics assumes the dynamo is maintained by interaction of large- and small-scale magnetic fields, with magnetic diffusion playing a critical role at the small scales (Soward, 1991). This basic assumption, which is supported by the results of many recent numerical dynamo models (Glatzmaier and Roberts, 1996; Kuang and Bloxham, 1997; Kageyama and Sato, 1997; Christensen et al., 1999; Olson et al., 1999; Kono et al., 2000a) leads to a very different balance of terms than implied by frozen flux, one that is more appropriate for interpreting long-term average magnetic field models.

To illustrate this difference, consider the lowest-order balance of terms for a nearly steady, nearly axisymmetric dynamo in the mean field electrodynamical approximation (Moffatt, 1978; Soward, 1991). The radial magnetic field and the tangential velocity are first separated into (steady) axisymmetric and nonaxisymmetric parts

$$B_r = B(r, \theta) + b(r, \theta, \phi, t), \tag{21}$$

$$\mathbf{U}_H = \mathbf{U}(r, \theta) + \mathbf{u}(r, \theta, \phi, t). \tag{22}$$

Substituting Eqs (21) and (22) into Eq. (19) and averaging in longitude ϕ and time t yields separate equations for the axisymmetric (zonal) magnetic field

$$(\mathbf{U} \cdot \boldsymbol{\nabla}_H)B + \langle (\mathbf{u} \cdot \boldsymbol{\nabla}_H)b \rangle + B(\boldsymbol{\nabla}_H \cdot \mathbf{U}) + \langle b(\boldsymbol{\nabla}_H \cdot \mathbf{u}) \rangle = \frac{\lambda}{r}\nabla^2(rB) \tag{23}$$

and for the nonaxisymmetric (nonzonal) field

$$\frac{\partial b}{\partial t} + (\mathbf{U} \cdot \boldsymbol{\nabla}_H)b + (\mathbf{u} \cdot \boldsymbol{\nabla}_H)B + B(\boldsymbol{\nabla}_H \cdot \mathbf{u}) + b(\boldsymbol{\nabla}_H \cdot \mathbf{U}) = \frac{\lambda}{r}\nabla^2(rb), \tag{24}$$

where $\langle \rangle$ denotes an average over ϕ and t.

Equations (23) and (24) are made dimensionless by introducing a velocity scale U and separate length scales L for the axisymmetric and l for the nonaxisymmetric variables. The dimensionless version of Eq. (24) is then

$$\epsilon \left[\frac{\partial b}{\partial t} + (\mathbf{U} \cdot \boldsymbol{\nabla}_H)b \right] + \delta(\mathbf{u} \cdot \boldsymbol{\nabla}_H)B + B(\boldsymbol{\nabla}_H \cdot \mathbf{u}) + \epsilon\delta(\boldsymbol{\nabla}_H \cdot \mathbf{U})b = \frac{\epsilon}{\delta Rm}\frac{1}{r}\nabla^2(rb), \tag{25}$$

where

$$\delta = l/L, \tag{26}$$

Rm is the magnetic Reynolds number defined in Eq. (20), and

$$\epsilon = |b|/|B| \tag{27}$$

is the ratio of nonaxisymmetric to axisymmetric field strengths. The mean field approximation assumes $\epsilon, \delta \ll 1$ and also $\epsilon \simeq \delta Rm$. To leading order in ϵ and δ, Eq. (25) reduces to (now in terms of dimensional variables)

$$B(\mathbf{\nabla}_H \cdot \mathbf{u}) = \frac{\lambda}{r} \nabla^2 (rb). \tag{28}$$

Projecting Eq. (28) onto the the CMB $r = r_c$ gives

$$Bw' \simeq \lambda \nabla_H^2 b, \tag{29}$$

where w' is the fluid upwelling below the CMB. According to this result, tangential magnetic field diffusion and upwellings are correlated in dynamos with mean field electrodynamic effects. This particular correlation is evident in many numerical dynamos, such as those shown in Figures 7, 8, and 15, for example. The correlation suggests a particularly straightforward way to interpret the nonaxisymmetric parts of the time-averaged historic geomagnetic and the paleomagnetic field on the CMB in terms of residual fluid upwellings and downwellings in the outer core.

A number of models of the core upwelling pattern have already been constructed by applying the frozen flux technique to the historical geomagnetic field and its secular variation. Unfortunately, there is little consistency between these models, since the upwelling planforms inferred with frozen flux depend strongly on which closure constraints chosen, that is, on whether the tangential flow is assumed to be geostrophic, steady, or toroidal. For example, the geostrophic constraint gives a prominent upwelling beneath the equatorial Indian Ocean and a downwelling beneath the equatorial East Pacific (Bloxham and Jackson, 1991), but no prominent upwelling structure beneath the Central Pacific. The steady flow constraint shows an upwelling beneath the Western Pacific but little structure in the Central Pacific (Voorhies, 1986, 1995; Bloxham and Jackson, 1991). This constraint locates most of the upwelling and downwelling structure in the Atlantic hemisphere, and relatively little in the Pacific hemisphere. The toroidal constraint assumes a priori there are no upwellings.

Consider, for example, the high field intensity beneath East Asia and Australia in the models of the time-averaged geomagnetic field shown in Figures 10–14. The interpretation of this structure is, according to Eq. (29), that it represents a fluid downwelling, roughly symmetric with respect to the equator. Its location is close to but slightly east of the high CMB heat flow region implied by the seismic structure of D″ shown in Figure 2. This is the interpretation of the high field intensity sectors that is consistent with the numerical dynamo models.

10. Mantle controls on geomagnetic polarity reversals

Since the mantle controls the rate of heat loss from the core, we expect that the thermal interaction of the core and mantle influences the time variability of the geomagnetic field, in addition to the time average properties of the field discussed in the previous sections.

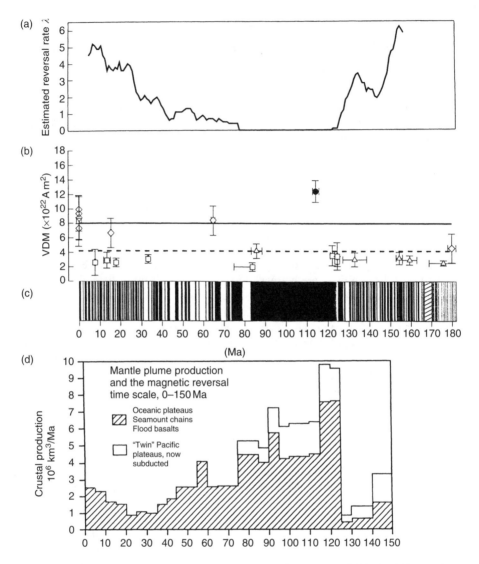

Figure 14 Comparison of long-term variations in the paleomagnetic field, 0–180 Ma and mantle plume activity. From top to bottom: (a) polarity reversal rate in Ma^{-1}; (b) virtual dipole moment (VDM) derived from paleointensity measurements; (c) polarity time scale, constrained by sea floor magnetic anomalies (black = normal polarity) and land data (gray = normal polarity). (Reprinted with permission from *Science* **291**, Tarduno, J. A., Cottrell, R. D. and Smirnov, A. V., "High geomagnetic intensity during the mid-Cretaceous from Thellier analyses of single plagioclase crystals," 1779–1782, figure 4. Copyright (2001) American Association for the Advancement of Science.) (d) Mantle plume activity 0–150 Ma as measured by crust production at hotspots. (Reprinted from *Earth Planet. Sci. Lett.* **107**, Larson, R. and Olson, P., "Mantle plumes control magnetic reversal frequency," 437–447. figure 1. Copyright (1991) with permission from Elsevier Science.)

Polarity reversals are the most significant element of geomagnetic variability, and the evidence suggests that the mantle may well exert some control on both the frequency of polarity reversals and the nature of the transition field during reversals.

10.1. Reversal frequency

The paleomagnetic record shows that the main dipole field has reversed its polarity several hundred times within the last 160 Ma. The average rate of reversal for this entire time period is about two polarity reversals per million year (Merrill *et al.*, 1996). However, when calculated over shorter intervals, the reversal frequency changes considerably with time. As shown in Figure 14, the reversal frequency has varied from about $4.5\,\mathrm{Ma}^{-1}$ at 160 Ma, then decreased to zero from 120 to 80 Ma, and has since increased to about $4.5\,\mathrm{Ma}^{-1}$ at present (McFadden and Merrill, 2000). The time interval without reversals, between about 118 and 83 Ma, is the Cretaceous Superchron. For a long time it was thought that the field intensity was lower than average during the Cretaceous Superchron. However, some recent data shown in Figure 14 indicate the field intensity was comparable to the average or even higher than average at that time.

There are several ways to interpret the changes in reversal frequency, each with its own implications for core–mantle interaction. One interpretation holds that the reversal frequency changes discontinuously from high values, such as in the recent past, to zero in the superchrons (Gallet and Hulot, 1997). This would imply that core–mantle interaction vacillates between two regimes, one characterized by uniformly frequent reversals, the other by no reversals. Alternatively, McFadden and Merrill (2000) have argued the data better support a model with gradual and continuous changes in reversal frequency. The implication of this interpretation is that the strength of core–mantle interaction is more cyclical in character, without abrupt changes from one preferred state to another.

In either case, the time scale of variation in Figure 14 is far longer than any time scale we can identify with fluid dynamical processes in the outer core. It is comparable, however, to the time scales of mantle flow. For this reason, the change in reversal frequency is often attributed to changes in the amplitude and pattern of heat flow on the CMB that accompany time-dependent mantle convection (Vogt, 1975; Larson and Olson, 1991). Some support for this interpretation comes from the relationship between reversal frequency and mantle plume activity, inferred from the rate of crustal formation at hotspots, and shown in Figure 14.

The question of how reversal rates are affected by conditions on the CMB is exceedingly difficult to answer, given our sketchy understanding of the cause of individual polarity reversals. Even so, it is possible to test the sensitivity of a particular dynamo model to different boundary conditions, and determine in a relative sense which boundary conditions promote reversals and which do not.

Figure 15 shows results of a suite of eight numerical dynamo models by Glatzmaier *et al.* (1999), calculated using different heat flux patterns on the outer boundary. In each case, the boundary heat flow pattern, the history of the dipole moment, dipole latitude, and dipole trajectory is shown. Polarity reversals are most frequent for cases 15(a) and (c). The boundary heat flux in case 15(a) lacks hemispherical symmetry, and this induces extreme time variability in the pole position. Case 15(c) has maximum boundary heat flux at the equator and minimum at the poles, and this pattern also results in frequent reversals, despite its equatorial symmetry. In contrast, case 15(e), with boundary heat flux maxima at both poles and at the equator,

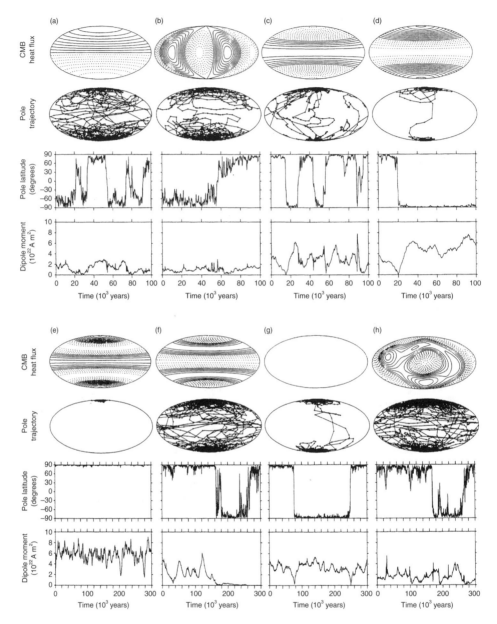

Figure 15 Summary of eight numerical dynamo simulations with different patterns of heat flow imposed at the outer boundary. Shown for each case are: first row, boundary heat flow pattern (solid contours = positive heat flow perturbations, broken contours = negative heat flow perturbations, both relative to the mean heat flow); second row, trajectory of south magnetic pole of the dipole magnetic field component over time; third row, south magnetic pole latitude versus time; fourth row, dipole moment versus time. (Reprinted with permission from Glatzmaier, G. A., Coe, R. C., Hongre, L., and Roberts, P. H., "The role of the Earth's mantle in controlling the frequency of geomagnetic reversals," *Nature* **401**, 885–890, figure 1 (1999).)

results in virtually no pole motion and no reversals. Case 15(c) is particularly significant, since this pattern of boundary heat flux approximates the internal heat transport pattern found in convection-driven dynamo models using simpler, homogeneous boundary conditions (Glatzmaier and Roberts, 1996; Kuang and Bloxham, 1997; Christensen *et al.*, 1999).

The calculations shown in Figure 15 have relatively low spatial resolution and, perhaps more importantly, rely on hyperdiffusivity for numerical convergence. Since both of these likely affect the polarity reversal rate, it is only the relative frequency of reversals that is physically meaningful here. Even with this limitation, several general inferences have been drawn from the Glatzmaier *et al.* (1999) calculations. First, the relative reversal rate is linked to the motion of the dipole. Motion of the dipole axis is a measure of time variability in the both the field and the flow. Second, boundary heat flux patterns that match the internal heat transport pattern for convection in a rotating sphere result in low variability flow and stable, high-intensity dipole fields, whereas boundary heat flux patterns that are incompatible with the internal heat transport pattern lead to higher variability, less stable and less intense dipole fields and more frequent reversals.

Figure 15(h) uses a heat flux pattern proportional to the D'' layer seismic structure, similar to Figure 3. The seismic tomography boundary heat flow pattern produces reversals at about the same frequency as the uniformly heated case 15(g), although the dipole is weaker and far less stable than with uniform boundary heating. The relatively large dipole variation in the tomographic heat flux case 15(h) results from the combination of zonal and azimuthal heat flow variation implied by the seismic tomography structure. As discussed earlier, zonal heat flow variations promote azimuthal flow, in the form of thermal winds. Instead of becoming locked to the mantle, the convection pattern continues to drift westward with the thermal wind in case 15(h), and is perturbed as it passes beneath the high heat flow sectors. Thermal winds produce additional time variation in the magnetic field, through the ω effect. By exciting thermal winds and modulating the strength of the drifting convection, the heterogeneous heating introduces additional time variability in both the convection and the magnetic field. Case 15(h) demonstrates the subtle balances that come into play with heterogeneous boundary heating. Here, the azimuthal heat flow variation fails to lock the flow pattern to the mantle structure. Instead, the combination of zonal and azimuthal heat flow variations implied by the D'' layer seismic structure tends to increase the time variability of the convection and promotes polarity reversals.

Although it is hazardous to make specific comparisons between numerical dynamo models and geomagnetic field behavior, the results in Figure 15 suggest a general interpretation of the long-term evolution in reversal frequency shown in Figure 14. The easiest way to increase the reversal frequency is to alter the CMB heat flow pattern in a way that excites thermal winds, particularly at high latitudes, where the dipole is located most of the time.

10.2. Reversal paths

Another manifestation of core–mantle coupling is the tendency for preferred states of the transition field during polarity changes. Paleomagnetic studies have compiled large datasets on transition field behavior (see Merrill and McFadden, 1999 and references contained therein). The general conclusion from these studies is that reversals differ from one another. There are certainly several, and possibly many ways in which the geomagnetic field changes its polarity. Even so, there are some broadly defined patterns in reversal behavior to suggest some influence of the mantle.

(a)

(b)

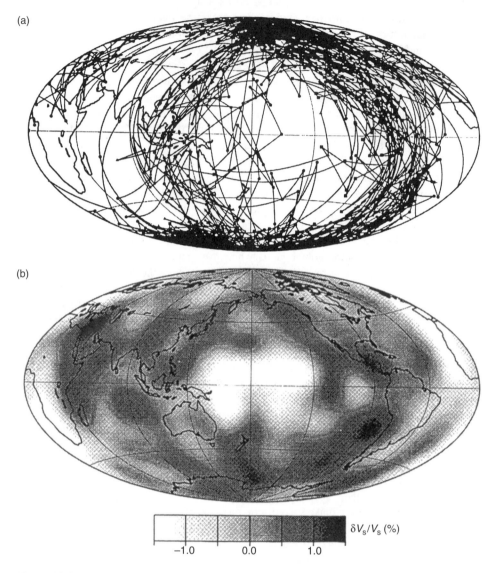

$\delta V_s/V_s$ (%)

−1.0   0.0   1.0

Figure 16 Comparison of transitional virtual geomagnetic pole paths VGP from 0 to 12 Ma compiled
by Laj *et al.* (1991) and D″ layer seismic shear velocity structure from Su *et al.* (1994).
The pole paths compiled by Laj *et al.* cluster along the high-velocity regions in the D″
layer. (Reprinted with permission from Aurnou, J. M., Buttles, J. L., Neumann, G. A., and
Olson, P., "Electromagnetic core–mantle coupling and reversal rates," *Geophys. Res. Lett.*
23, 270–273, figure 2 (1996).)

The best evidence for mantle influence on individual reversals comes from longitudinal
biases in the transition field. This evidence includes: (i) some tendency for the virtual geomagnetic pole (VGP) to dwell in certain regions during transition (Hoffmann, 1996, 2000);
and (ii) a statistical preference for the transitional VGP to prefer either of two longitudinal
bands, one beneath the Americas, the other beneath East Asia. The existence of preferred

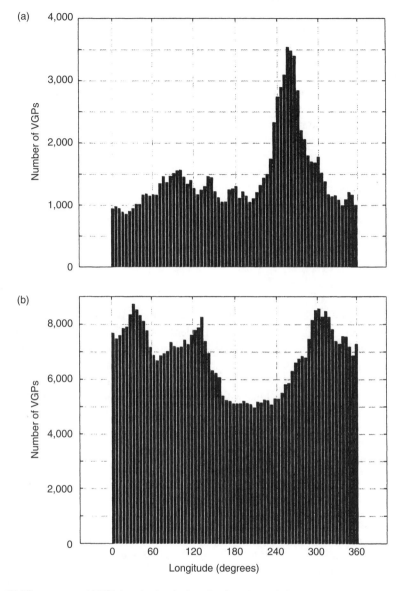

Figure 17 Histograms of VGP longitudes during the first (a) and the second (b) magnetic polarity reversals from the numerical dynamo model shown in Figure 15(h). (Reprinted with permission from Coe, R. S., Hongre, L., and Glatzmaier, G. A., "An examination of simulated geomagnetic reversals from a palaeomagnetic perspective," *Phil. Trans. R. Soc. Lond.* A **358**, 991–1000, figures 16 and 17 (2000).)

longitudes for the transition field has been debated for a decade (Laj *et al.*, 1991; Prevot and Camps, 1993; Love, 1998). Recent efforts to delineate preferred longitudes using large datasets of transition VGPs seem to indicate there is a statistical preference for the pole to dwell in two sectors, one centered near 60°W, the other centered near 90°E, as illustrated in Figure 16.

There are many unanswered questions regarding preferred transition field longitudes. In the first place, not everyone agrees the preference for certain longitudes is genuine (see Dormy *et al.*, 2000 for a discussion of this point). Second, even if the effect is genuine, there is little agreement on its cause. It could be a thermal effect, in which the transition field is biased the same way the time average field during a polarity chron could be biased by heterogeneous CMB heat flow. Figure 17 shows VGP density calculated by Coe *et al.* (2000) from transition fields during the two polarity reversals captured in a numerical dynamo with tomographic thermal boundary conditions (case (h) in Figure 15). Although the two reversals seem to have occurred through different mechanisms, and are dissimilar in many respects, there is a weak, statistical correlation between transition field VGP density and the boundary heating pattern. In particular, the VGP density is relatively high at longitudes where the boundary heat flow is high, and conversely, the VGP density is relatively low where the boundary heat flow is low. This correlation can be explained by the tendency for the radial field to concentrate in downwellings beneath regions with high boundary heat flow, as discussed in the previous sections. However, the effect is rather weak. The model results indicate the transitional VGP paths are controlled more by MHD processes in the core at the time of polarity reversal than by conditions at the CMB.

11. Summary

The case for long-term thermal interaction between the core and the mantle continues to strengthen with time. Even so, the best evidence for it remains largely indirect, and can be interpreted in several different ways. In this situation, multiple working hypotheses are to be encouraged, particularly those hypotheses that can be explored with model studies and tested against observations. This review has highlighted several important objectives that, once met, would go a long way to clarifying the nature of core–mantle thermal interaction.

One important objective is to better constrain the total heat flow at the CMB and its spatial variation. This is not an easy task, as it requires more certain interpretation of the observed seismic heterogeneity in the D'' layer than currently exists. A second and equally important objective is to better delineate the nonaxisymmetric structure of the paleomagnetic field. It seems that a concerted effort in this direction could establish a definitive paleomagnetic field model for the last 5 Ma or so, which would serve as the basis for comparison with numerical models of the geodynamo that incorporate a heterogeneous CMB. Lastly, there are two important theoretical questions on how the geodynamo responds to CMB thermal heterogeneity for which we have very incomplete answers to, but could be answered by systematic numerical study. The first issue is to determine what factors control the longitudinal phase relationship between time-averaged observable properties of the dynamo (such as the time average nonaxisymmetric magnetic field) and the pattern of variable boundary heating. The second theoretical issue is to isolate what part of CMB thermal heterogeneity most affects the stability of the geodynamo in time, by altering the amplitude of secular variation, and frequency of excursions and polarity reversals. Important steps toward answering these questions have already been made, but there is certainly much room for future progress.

References

Aurnou, J. M. and Olson, P. L., "Experiments on Rayleigh–Benard convection, magnetoconvection, and rotating magnetoconvection in liquid gallium," *J. Fluid Mech.* **430**, 283–307 (2001).

Aurnou, J. M., Buttles, J. L., Neumann, G. A., and Olson, P., "Electromagnetic core–mantle coupling and reversal paths," *Geophys. Res. Lett.* **23**, 270–273 (1996).

Bloxham, J., "The effect of thermal core–mantle interactions on the paleomagnetic secular variation," *Phil. Trans. R. Soc. Lond.* A **358**, 1171–1179 (2000).

Bloxham, J. and Gubbins, D., "Thermal core–mantle interactions," *Nature* **325**, 511–513 (1987).

Bloxham, J. and Jackson, A., "Lateral temperature variations at the core–mantle boundary deduced from the magnetic field," *Geophys. Res. Lett.* **17**, 1997–2000 (1990).

Bloxham, J. and Jackson, A., "Fluid flow near the surface of Earth's outer core," *Rev. Geophys.* **29**, 97–120 (1991).

Bloxham, J. and Jackson, A., "Time dependent mapping of the geomagnetic field at the core–mantle boundary," *J. Geophys. Res.* **97**, 19357–19564 (1992).

Boehler, R., "High-pressure experiments and the phase diagram of lower mantle and core materials," *Rev. Geophys.* **38**, 221–245 (2000).

Bolton, E. W. and Sayler, B. S., "The influence of lateral variations of thermal boundary conditions on core convection: numerical and laboratory experiments," *Geophys. Astrophys. Fluid Dynam.* **60**, 369–370 (1991).

Buffett, B. A., "Earth's core and the geodynamo," *Science* **288**, 2007–2012 (2000).

Buffett, B. A., Huppert, H. E., Lister, J. R., and Woods, A. W., "Analytical model for solidification of the Earth's core," *Nature* **356**, 329–331 (1992).

Buffett, B. A., Huppert, H. E., Lister, J. R., and Woods, A. W., "On the thermal evolution of the Earth's core," *J. Geophys. Res.* **101**, 7989–8006 (1996).

Buffett, B. A., Garnero, E. J., and Jeanloz, R., "Sediments at the top of Earth's core," *Science* **290**, 1338–1342 (2000).

Busse, F. H. and Carrigan, C. R., "Laboratory simulation of thermal convection in rotating planets and stars," *Science* **191**, 81–83 (1976).

Calderwood, A. R., "The distribution of U, Th, and K in the Earth's crust, lithosphere, mantle and core: constraints from an elemental mass balance model and the present day heat flux." In: *SEDI 2000*, the 7th Symposium of Study of the Earths Deep Interior, Exeter, UK, Abstract S9.4 (2000).

Christensen, U., Olson, P., and Glatzmaier, G. A., "Numerical modeling of the geodynamo: A systematic parameter study," *Geophys. J. Int.* **25**, 1565–1568 (1999).

Coe, R. S., Hongre, L., and Glatzmaier, G. A., "An examination of simulated geomagnetic reversals from a palaeomagnetic perspective," *Phil. Trans. R. Soc. Lond.* A **358**, 1141–1170 (2000).

Constable, C. G., Johnson, C. L., and Lund, S. P., "Global geomagnetic field models for the past 3000 years: transient or permanent flux lobes?" *Phil. Trans. R. Soc. Lond.* A **358**, 991–1008 (2000).

Dormy, E., Valet, J.-P., and Courtillot, V., "Numerical models of the geodynamo and observational constraints," *Geochem. Geophys. Geosys.* VI, paper 2000 GC000062 (2000).

Gallet, Y. and Hulot, G., "Stationary and nonstationary behavior within the geomagnetic polarity time scale," *Geophys. Res. Lett.* **24**, 1875–1878 (1997).

Garnero, E. J., Revenaugh, J., Williams, Q., Lay, T., and Kellogg, L. H., "Ultralow velocity zone at the core–mantle boundary," In: *The Core–mantle Boundary Region* (Eds M. Gurnis, M. E. Wysession, E. Knittle, and B. A. Buffet) Geodynamics Series **28**, pp. 319–334, AGU (1998).

Gibbons, S. J. and Gubbins, D., "Convection in the Earth's core driven by lateral variations in the core–mantle boundary heat flux," *Geophys. J. Int.* **142**, 631–642 (2000).

Glatzmaier, G. A. and Roberts, P. H., "Rotation and magnetism of Earth's inner core," *Science* **274**, 1887–1891 (1996).

Glatzmaier, G. A., Coe, R. C., Hongre, L., and Roberts, P. H., "The role of the Earth's mantle in controlling the frequency of geomagnetic reversals," *Nature* **401**, 885–890 (1999).

Gubbins, D., "Interpreting the paleomagnetic field," In: *The Core–Mantle Boundary Region* (Eds M. Gurnis, M. E. Wysession, E. Knittle, and B. A. Buffett) pp. 167–182, AGU (1997).

Gubbins, D. and Bloxham, J., "Morphology of the geomagnetic field and implications for the geodynamo," *Nature* **325**, 509–511 (1987).

Gubbins, D. and Richards, M., "Coupling of the core dynamo and mantle: thermal or topographic?" *Geophys. Res. Lett.* **13**, 1521–1524 (1986).

Hart, J. E., Toomre, J., Deane, A. E., Hurlburt, N. E., Glatzmaier, G., Fichtl, G. h., Leslie, F., Fowlis, W. W., and Gilman, P. A., "Laboratory experiments on planetary and stellar convection performed on Spacelab 3," *Science* **234**, 61–64 (1986).

Hide, R. "On the earth's core–mantle interface," *Q. J. R. Met. Soc.* **96**, 579–590 (1970).

Jeanloz, R., "The nature of Earth's core," *Annu. Rev. Earth Planet. Sci.* **18**, 357–386 (1990).

Johnson, C. L. and Constable, C. G., "The time averaged geomagnetic field as recorded by lava flows over the past 5 Myr," *Geophys. J. Int.* **122**, 489–519 (1995).

Johnson, C. L. and Constable, C. G., "The time averaged geomagnetic field: global and regional biases for 0–5 Ma," *Geophys. J. Int.* **131**, 643–666 (1997).

Johnson, C. L. and Constable, C. G., "Persistently anomalous Pacific geomagnetic fields," *Geophys. Res. Lett.* **25**, 1011–1014 (1998).

Jones, G. M., "Thermal interaction of the core and the mantle and long-term behaviour of the geomagnetic field," *J. Geophys. Res.* **82**, 1703–1709 (1977).

Kageyama, A. and Sato, T., "Generation mechanism of a dipole field by a magnetohydrodynamical dynamo," *Phys. Rev. E* 4617–4626 (1997).

Kelly, P. and Gubbins, D., "The geomagnetic field over the past 5 million years," *Geophys. J. Int.* **128**, 315–330 (1997).

Kono, M., Sakuraba, A., and Ishida, M., "Dynamo simulation and palaeosecular variation models," *Phil. Trans. R. Soc. Lond.* A **358**, 1123–1139 (2000a).

Kono, M., Tanaka, H., and Tsunakawa, H., "Spherical harmonic analysis of paleomagnetic data: the case of linear mapping," *J. Geophys. Res.* **105**, 5817–5833 (2000b).

Kuang, W. and Bloxham, J., "An Earth-like numerical dynamo model," *Nature* **389**, 371–374 (1997).

Kuo, B.-Y., Garnero, E. J., and Lay, T., "Tomographic inversion of S-SKS times for shear velocity heterogeneity in D″: Degree 12 and hybrid models," *J. Geophys. Res.* **105**, 28138–28157 (2000).

Laio, A., Bernard, S., Chiarotti, G. L., Scandolo, S., and Tosatti, E., "Physics of iron at Earth's core conditions," *Science* **287**, 1027–1030 (2000).

Laj, C., Mazaud, A., Weeks, R., Fuller, M., and Herrero-Bervara, E., "Geomagnetic reversal paths," *Nature* **351**, 447 (1991).

Larson, R. and Olson, P., "Mantle plumes control magnetic reversal frequency," *Earth Planet Sci. Lett.* **107**, 437–447 (1991).

Li, X. and Romanowicz, B., "Global mantle shear velocity models developed using nonlinear asymptotic coupling theory," *J. Geophys. Res.* **101**, 22245–22272 (1996).

Lister, J. R. and Buffett, B. A., "The strength and efficiency of thermal and compositional convection in the geodynamo," *Phys. Earth Planet. Inter.* **91**, 17–30 (1995).

Lister, J. R. and Buffett, B. A., "Stratification of the outer core at the core–mantle boundary," *Phys. Earth Planet. Inter.* **105**, 5–19 (1998).

Loper, D. E., "The gravitationally powered dynamo," *Geophys. J. R. Astr. Soc.* **54**, 389–404 (1978a).

Loper, D. E., "Some thermal consequences of the gravitationally powered dynamo," *J. Geophys. Res.* **83**, 5961–5970 (1978b).

Loper, D. E. and Lay, T., "The core–mantle boundary region," *J. Geophys. Res.* **100**, 6397–6420 (1995).

Love, J. J., "Paleomagnetic volcanic data and geometric regularity of reversals and excursions," *J. Geophys. Res.* **103**, 12435–12452 (1998).

McFadden, P. L. and Merrill, R. T., "Evolution of the geomagnetic reversal rate since 160 Ma: is the process continuous?" *J. Geophys. Res.* **105**, 28455–28460 (2000).

Merrill, R. T. and McFadden, P. L., "Geomagnetic polarity transitions," *Rev. Geophys.* **37**, 201–266 (1999).

Merrill, R. T., McElhinny, M. W., and McFadden, P. L., *The Magnetic Field of the Earth: Paleomagnetism, the Core, and the Deep Mantle.* Academic Press, San Diego (1996).

Moffatt, H. K., *Magnetic Field Generation in Electrically Conducting Fluids.* Cambridge University Press, Cambridge (1978).

Olson, P. and Glatzmaier, G. A., "Magnetoconvection and thermal coupling of the Earth's core and mantle," *Phil. Trans. R. Soc. Lond.* A **354**, 1413-1424 (1996).

Olson, P., Christensen, U., and Glatzmaier, G. A., "Numerical modeling of the geodynamo: mechanisms of field generation and equilibration," *J. Geophys. Res.* **104**, 10383–10404 (1999).

Prevot, M. and Camps, P., "Absence of preferred longitudinal sectors for poles from volcanic records of geomagnetic reversals," *Nature* **366**, 53–57 (1993).

Roberts, P. H. and Gubbins, D., "Origin of the main field: kinematics," In: *Geomagnetism*, vol. 2 (Ed. J. A. Jacobs) pp. 184–249. Academic Press, London (1987).

Roberts, P. H. and Scott, S., "On analysis of secular variation, 1, A hydromagnetic constraint," *J. Geomag. Geoelectr.* **17**, 137–151 (1965).

Sarson, G. R., Jones, C. A., and Longbottom, A. W., "The influence of boundary region heterogenieties on the geodynamo," *Phys. Earth Planet. Inter.* **101**, 13–32 (1997).

Schubert, J., Turcotte, D., and Olson, P., *Mantle Convection in the Earth and Planets*. Cambridge University Press, New York (in press) (2001).

Soward, A. M., "The Earth's dynamo," *Geophys. Astrophys. Fluid Dynam.* **62**, 191–209 (1991).

Stevenson, D. J., "Limits on lateral density variations in the Earth's outer core," *Geophys. J. R. Astr. Soc.* **88**, 311–319 (1987).

Stixrude, L., Wasserman, E., and Cohen, R. E., "Composition and temperature of Earth's inner core," *J. Geophys. Res.* **102**, 24729–24739 (1997).

Su, W. J., Woodward, R. L., and Dziewonski, A. M., "Degree 12 model of shear velocity heterogeneity in the mantle," *J. Geophys. Res.* **99**, 6945–6980 (1994).

Sumita, I. and Olson, P., "A laboratory model for convection in Earth's core driven by a thermally heterogeneous mantle," *Science* **286**, 1547–1549 (1999).

Sun, Z.-P., Schubert, G., and Glatzmaier, G. A., "Numerical simulations of thermal convection in a rapidly rotating spherical shell cooled inhomogeneously from above," *Geophys. Astrophys. Fluid Dynam.* **75**, 199–226 (1994).

Tackley, P. J., "Mantle convection and plate tectonics: towards an integrated physical and chemical theory," *Science* **288**, 2002–2007 (2000).

Tackley, P. J., Stevenson, D. J., Glatzmaier, G. A., and Schubert, G., "Effects of multiple phase transitions in a three-dimensional spherical model of convection in Earth's mantle," *J. Geophys. Res.* **99**, 15877–15901 (1994).

Tarduno, J. A., Cottrell, R. D., and Smirnov, A. V., "High geomagnetic intensity during the mid-Cretaceous from Thellier analyses of single plagioclase crystals," *Science* **291**, 1779–1782 (2001).

van der Hilst, R. D. and Kárason, H., "Compositional heterogeneity in the bottom 1000 kilometers of Earth's mantle: toward a hybrid convection model," *Science* **283**, 1885–1888 (1999).

Vogt, P. R., "Changes in geomagnetic reversal frequency at times of tectonic change: evidence for coupling between core and upper mantle processes," *Earth Planet Sci. Lett.* **25**, 313–321 (1975).

Voorhies, C. V., "Steady flows at the top of Earth's core derived from geomagnetic field models," *J. Geophys. Res.* **91**, 12444–12466 (1986).

Voorhies, C. V., "Time-varying flows at the top of Earth's core derived from definitive geomagnetic reference models," *J. Geophys. Res.* **100**, 10029–10039 (1995).

Whaler, K. A., "Geomagnetic evidence for fluid upwelling at the core–mantle boundary," *Geophys. J. R. Astr. Soc.* **86**, 563–588 (1986).

Williams, Q. and Garnero, E. J., "Seismic evidence for partial melt at the base of the Earth's mantle," *Science* **273**, 1528–1529 (1996).

Wysession, M. E., Langenhorst, A., Fouch, M. J., Fischer, K. M., Al-Eqabi, G. I., Shore, P. J., and Clarke, T. J., "Lateral variations in compressional/shear velocities at the base of the mantle," *Science* **284**, 120–125 (1999).

Yoshida, S. and Hamano, Y., "Fluid motion of the outer core in response to a temperature heterogeneity at the core–mantle boundary and its dynamo action," *J. Geomag. Geoelectr.* **45**, 1497–1516 (1993).

Yoshida, S. and Shudo, E., "Linear response of the outer core fluid to the thermal heterogeneity on the core–mantle boundary," In: *SEDI 2000*, the 7th Symposium on Studies of the Earth's Deep Interior, Exeter, UK, Abstract vol. S3.10 (2000).

Yuen, D. A., Cadek, O., Chopelas, A., and Matyska, C., "Geophysical inferences of thermal–chemical structures in the lower mantle," *Geophys. Res. Lett.* **20**, 899–902 (1993).

Yuen, D. A., Cadek, O., van Keken, P., Reutleer, D. M., Kyvalova, H., and Schroeder, B. A., "Combined results from mineral physics, tomography and mantle convection and their implications on global geodynamics," In *Seismic Modelling of Earth Structure*, (Eds E. Boschi, G. Ekstrom, and A. Morelli) pp. 463–505. Instituto Nazionale de Geofisica, Rome (1996).

Zhang, K. and Gubbins, D., "On convection in the earth's core driven by lateral temperature variations in the lower mantle," *Geophys. J. Int.* **108**, 247–255 (1992).

Zhang, K. and Gubbins, D., "Convection in a rotating spherical fluid shell with an inhomogeneous temperature boundary condition at infinite Prandtl number," *J. Fluid Mech.* **250**, 209–232 (1993).

Zhang, K. and Gubbins, D., "Convection in a rotating spherical shell with an inhomogeneous temperature boundary condition at finite Prandtl number," *Phys. Fluids* **8**, 1141–1148 (1996).

2 Convection and the lower mantle

Gerald Schubert

Department of Earth and Space Sciences, Institute of Geophysics and Planetary Physics, University of California, Los Angeles, CA 90095-1567, USA

An assessment is made of the proposal that an isolated, compositionally dense, heat source rich layer exists in the bottom half of the lower mantle. The Earth's heat budget is reviewed and it is concluded that though the fraction of the Earth's heat loss attributable to mantle radioactivity is uncertain, an isolated mantle reservoir, rich in radioactive elements, is not required to balance the thermal budget. The balance does require either a large flow of heat from the core, or a large contribution from secular cooling of the mantle, or a mantle-wide concentration of radioactive elements larger than predicted by geochemical and cosmochemical models. It is argued that the disappearance of descending slabs in seismic tomographic images of the lower half of the lower mantle is not necessarily due to ponding of the slabs on the top of a compositionally dense lower mantle layer. Other explanations are equally plausible including the gradual reabsorption of descending slabs into the background mantle circulation. Most importantly, it is emphasized that a compositionally dense, heat source rich layer in the bottom half of the lower mantle cannot be hidden from seismic view by offsetting effects of temperature and composition on seismic velocity. The thermal boundary layer at the top of the lower layer would produce a distinct triplication in short-period P waves; seismic data show no evidence of P-wave triplications due to any mid-mantle thermal boundary layers. The thermal boundary layer at the top of the lower layer would also inhibit cooling of the lower layer and make it hot enough to be partially molten, in disagreement with seismic evidence for the solidity of the lower mantle. It is concluded that there is no isolated, compositionally dense, heat source rich layer in the bottom half of the Earth's lower mantle.

1. Introduction

Even after three decades of intensive study of the mantle and its dynamics there are still basic uncertainties about the nature of mantle convection. In the past, the debate about the nature of mantle convection centered on whether there was whole-mantle convection or two-layer mantle convection, with the seismic discontinuity at 660 km depth forming the boundary between separately convecting upper and lower mantle regions. Much of this debate was driven by geochemical arguments that articulated the need for an isolated reservoir of primordial material in the mantle (Albarede and van der Hilst, 1999). Seismic tomographic images of the mantle leave little doubt that at least some descending slabs pass through 660 km depth, enter the lower mantle, and sink all the way to the core–mantle boundary (CMB) or to its close proximity (Grand *et al.*, 1997; van der Hilst *et al.*, 1997). This rules out the model of separately convecting upper and lower mantle regions but it does not establish the validity of whole mantle convection because some slabs seem to be deflected horizontally at 660 km depth (Fukao *et al.*, 1992), while others that have entered the lower mantle seem to "disappear" somewhere in the mid-lower mantle (van der Hilst and Karason, 1999). The nature of mantle convection is obviously more complex than the unimpeded sinking of all descending slabs to the very bottom of the mantle.

One possibility for the additional complexity in mantle structure and dynamics is the model proposed by Kellogg *et al.* (1999) and van der Hilst and Karason (1999). Partly on the basis of the disappearance of linear slab signatures below mid-lower mantle depths in seismic tomographic models and partly on the basis of problems (discussed below) in understanding Earth's heat flow, these authors hypothesized the existence of an essentially isolated, variable thickness, compositionally distinct region in the lower half of the lower mantle containing a primordial concentration of radiogenic heat sources. Slabs descending into the bottom half of the lower mantle would penetrate to different depths, depending on their negative buoyancy, by pushing aside the material of the bottom lower mantle. Some slabs might descend to the proximity of the CMB, while other slabs might be turned aside near mid-lower mantle depths as sketched in Figure 1. The inability of some slabs to penetrate below middle lower mantle depths could explain the disappearance of the linear slab signal in the seismic tomography. Thermal and compositional effects on density and seismic velocity could offset each other to reduce the seismic detectability of the deep lower mantle layer.

This model is a variation on the classical two-layer model of mantle convection and is motivated by some of the same geochemical considerations. These include: (a) characteristic

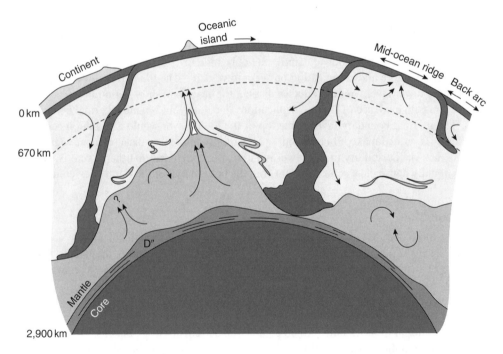

Figure 1 A sketch of what convection might look like in a model of the mantle containing an isolated, compositionally dense, heat source rich, thick lower layer. Descending slabs push aside the lower layer material and create an undulating topography at the top of the lower layer. (Reprinted with permission from Kellogg, L. H., Hager, B. H. and van der Hilst, R. D., "Compositional stratification in the deep mantle," *Science* **283**, 1883–1884 (1999). Copyright 1999, American Association for the Advancement of Science.)

isotopic signatures of mid-ocean ridge basalts (MORBs) and ocean island basalts (OIBs) that suggest distinct mantle reservoirs (Hofmann, 1997); (b) strontium and neodymium isotope ratios of the crust and MORB that suggest the crust was extracted from substantially less than the entire mantle (Jacobsen and Wasserburg, 1979; Wasserburg and DePaolo, 1979; O'Nions *et al.*, 1979); (c) the ^{40}Ar content of the atmosphere which requires less than 100% degassing of the mantle (Allegre *et al.*, 1983, 1996; Turcotte and Schubert, 1988); and (d) the relatively high (nonradiogenic) ^{3}He/^{4}He ratios in OIB, which also indicate that the mantle has not been completely outgassed. Instead of the interface between the layers coinciding with the boundary between the upper and lower mantle at 660 km depth, the Kellogg *et al.* (1999) and van der Hilst and Karason (1999) model has the seismically camouflaged interface buried deep in the lower mantle. In the following, we discuss the bases and consequences of the Kellogg *et al.* (1999) and van der Hilst and Karason (1999) model of the lower mantle and assess its relevance to the real Earth. The viability of the model is tied to the fate of descending slabs in the lower mantle and this depends critically on both slab buoyancy and rheology. We point out that even if the bulk of the putative deep lower mantle layer is seismically similar to that of the lower mantle just above it, there will exist a prominent thermal boundary layer at the interface that would be seismically detectable if the deep layer actually existed (Vidale *et al.*, 2001). A deep lower mantle layer with a high concentration of radiogenic heat sources cannot be hidden from seismic view.

2. Heat flow considerations

One of the main arguments used by Kellogg *et al.* (1999) to hypothesize the existence of a deep, isolated, chemically distinct lower mantle layer is the supposed need for a mantle reservoir rich in radioactivity to account for the heat flow through the Earth's surface. Is such a reservoir really needed? What do we know about the heat flow from the Earth and its source in the Earth's interior? It is generally agreed that about 44 TW of heat flows through the Earth's surface although that number is not measured since we cannot measure the heat loss in the vicinity of mid-ocean ridges and there are no heat flow measurements over large areas of the continents (e.g. Pollack *et al.*, 1993; Schubert *et al.*, 2001).

About 7.5 TW originates from radiogenic heating in the continental crust (Schubert *et al.*, 2001); this value is only an approximation because the concentration of radioactivity in the lower crust is uncertain and we must estimate the amount of heat entering the base of the crust from the underlying mantle. If we adopt this value for the radiogenic heat production of the continental crust, then 36.5 TW of heat flows out of the Earth's mantle. A fraction of this heat originates in the core, a portion is produced by radioactive decay in the mantle, and the rest derives from the secular cooling of the mantle. The main question, of course, is what are these relative contributions to the heat flow? Kellogg *et al.* (1999) accept a geochemically based estimate of radiogenic heat production in the upper mantle (U = 7 ppb, Th/U = 2.5, K/U = 10^4) and calculate that only about 6 TW could be produced by this concentration of radioactivity distributed throughout the mantle. They hypothesize that there is a high concentration of radioactivity (U = 25.6 ppb, Th/U = 4, K/U = 10^4) in a primordial deep lower mantle layer. This amount of radioactivity produces almost 25 TW of heat if it is distributed throughout the mantle and less if it is just in the lower layer (reduce 25 TW by the fraction of the mantle mass in the deep primordial lower layer). Since 36.5 TW must be accounted for, it is clear that in the Kellogg *et al.* (1999) model a substantial fraction of the Earth's heat loss must be attributed to heat lost from the core and/or the secular cooling of the mantle. The low concentration of radioactivity in the upper mantle assumed by Kellogg *et al.* (1999) is inferred from measured concentrations of radioactivity in MORBs (U $= 71.1$ ppb, Th $= 187.1$ ppb, K $= 883.7$ ppm, Th/U $= 2.63$, K/U $= 4.7 \times 10^3$) (Hofmann, 1988). However, the inference of radiogenic element concentrations in the MORB source region from concentrations of radiogenic elements measured in oceanic basalts is uncertain. It involves knowing how radiogenic elements are partitioned and concentrated in the melts that eventually solidify into MORB and the melt history of the basalts.

An alternative way to estimate the amount of radiogenic heating in the mantle is to estimate the contributions of core heat flow and mantle secular cooling to the 36.5 TW of mantle heat flow and attribute the difference to mantle radioactivity. Of course, these other quantities are also uncertain. Davies (1988) and Sleep (1990) calculated the mantle plume flux from known hotspot swells and equated this to the heat flow from the core; they obtained only about 2.4 TW of heat leaving the core. But Malamud and Turcotte (1999) derived a much larger value of core heat loss, 13.4 TW, from a model that hypothesizes the existence of mantle plumes that are not evident at the surface. The existence of the geodynamo provides only a lower bound on the heat flow from the core that is consistent with the estimates of Davies (1988) and Sleep (1990). Calderwood (2000) has argued that there could be a large amount of K in the core and that a large fraction of the 36.5 TW leaving the mantle is produced in the core. It is clear that the core heat flux estimates of Davies (1988) and Sleep (1990) provide only a lower bound on the heat flow from the core. Heat from the core probably warms mantle material near the CMB (cold slabs perhaps) that later rises through the mantle as part of the general circulation or in unidentified plumes and deposits heat at the base of the continents or at the base of old oceanic lithosphere. This heat is not accounted for by an inventory of hotspot plume flux. The core could be a larger source of heat than hitherto appreciated thereby reducing the amount of heat that needs to be accounted for by mantle radioactivity.

The contribution of secular cooling to the heat loss from the mantle is similarly uncertain. The Urey number Ur is a dimensionless parameter commonly used in geodynamics to specify the fraction of mantle heat flow derived from secular cooling (actually, the fraction is $1 - Ur$). Turcotte and Schubert (1982) and Schubert *et al.* (2001) use theoretical models to estimate that the Urey number is about 0.8. However, others have argued that Ur could be as small as 0.4, or that 60% of the mantle heat flow could be due to secular cooling. Lord Kelvin took the Urey number to be zero and prior to papers by Sharpe and Peltier (1979), Stevenson and Turner (1979), Davies (1980), Schubert *et al.* (1980), Stacey (1980) and Turcotte (1980) geodynamicists took $Ur = 1$. We do not know the actual value of Ur. With $Ur = 0.8$, values for mantle radiogenic heat production are U $= 31$ ppb, Th $= 124$ ppb, K $= 310$ ppm, Th/U $= 4$, K/U $= 10^4$ corresponding to a mantle radiogenic heat production of 29.5 TW and a mantle secular cooling contribution of 7 TW. Are these values of mantle radioactivity reasonable? McDonough and Sun (1995) propose that U $= 20.3$ ppb, Th $= 79.5$ ppb, K $= 240$ ppm, Th/U $= 3.92$, K/U $= 1.18 \times 10^4$ for the bulk silicate Earth (which includes the crust) based on concentrations of radiogenic elements in CI meteorites (U $= 7.4$ ppb, Th $= 79.5$ ppb, K $= 550$ ppm, Th/U $= 10.7$, K/U $= 6.9 \times 10^3$). With the bulk silicate Earth radiogenic element concentrations, radiogenic heat flow from the silicate Earth is about 20 TW (this includes the radiogenic heat produced in the continental crust). The main difficulty with assessing the relevance of this cosmochemical estimate of Earth's radioactivity is the uncertainty we have in determining which chondritic abundance best represents the Earth; there is a considerable range of values of radiogenic element abundances in different classes of chondrites as summarized in Table 1. There is also uncertainty associated with the issue of the degree of potassium depletion in the Earth (Table 1). Nevertheless, all the geochemical

Table 1 Chondritic compositions and heat production in an earth-size planet[a] (chondritic abundances are from Wasson and Kallemeyn, 1988)[b]

	Mg (mg/g)	*K* (µg/g)	*Th* (ng/g)	*U* (ng/g)	*Heat production* (TW)			
					K	*Th*	*U*	*Total*
CI-devol[c]	140	808	42	12	17	6.6	6.9	30
H	140	780	42	12	16	6.6	7.0	30
CV	145	310	60	17	6.5	9.5	10.0	26
EH	106	800	30	9	17	4.7	5.3	27
K depleted by a factor of 5								
CI-devol[c]	140	162	42	12	3.4	6.6	6.9	17
H	140	156	42	12	3.2	6.6	7.0	17
CV	145	62	60	17	1.3	9.5	10.0	21
EH	106	160	30	9	3.3	4.7	5.3	13
All K values set $= 160\,\mu g/g$								
CI-devol[c]	140	160	41.9	12	3.3	6.6	6.9	17
H	140	160	42	12	3.3	6.6	7.0	17
CV	145	160	60	17	3.3	9.5	10.0	23
EH	106	160	30	9	3.3	4.7	5.3	13

Notes
a Prepared by J. T. Wasson.
b See above.
c CI-devol entries are CI-chondrite abundances with volatiles removed (Mg arbitrarily set at 140 mg/g rather than 97 mg/g; other abundances adjusted by the same factor).

and cosmochemical estimates of heat from mantle radioactivity are significantly smaller than the estimate above based on the theoretically predicted value of the Urey number.

Whether we try to estimate the concentration of radiogenic elements in the mantle directly from geochemical data or indirectly from geophysical approaches it must be concluded that the concentration is uncertain. Nevertheless, there is no compelling reason to conclude that there must be an isolated region of the mantle with enhanced concentrations of the radiogenic elements.

3. Seismological considerations

One of the principal arguments used by van der Hilst and Karason (1999) to support the existence of a deep, isolated, chemically distinct lower mantle layer is the disappearance of the linear slab signal in seismic tomography at depths greater than about 1,500 km. Linear structures in tomographic images that are prominent at 1,250 km depth tend to disintegrate in the mid-lower mantle with only some parts of them connecting to the CMB beneath central America and east Asia. One explanation for this is that some slabs do not penetrate below mid-lower mantle depths possibly because they cannot enter the compositionally distinct deep lower mantle layer. However, if there were such a barrier, one would expect to see large-scale horizontal structure due to horizontal deflection of downwellings in the middle of the lower mantle (which one does not see) instead of near the CMB (which one does see). In addition, there are other possible explanations of this observation. It may be that slabs do tend to tend to lose their identity and merge into the background as they approach the bottom of the lower mantle as occurs with downwellings in numerical models of three-dimensional mantle convection (Bercovici *et al.*, 1989a,b). The loss of identification in seismic tomography may mean that slabs become difficult to detect seismically below mid-lower mantle depths even though they may penetrate to such depths. The cessation of earthquakes at about 660 km depth was once used as an argument against slab penetration into the lower mantle; just because slabs stopped revealing their presence through earthquakes did not mean that the slabs were not in the lower mantle. Descending slabs may undergo some rheological changes in the deep lower mantle that promote their reabsorption into the background mantle. Slab heating with increased depth of penetration reduces slab viscosity and enhances the equilibration of the slab with the surrounding mantle.

Other seismological data relevant to the composition of the lower mantle are the depth variations of the ratios

$$R_{S/P} \equiv \frac{\partial(\ln v_S)}{\partial(\ln v_P)}, \quad R_{\Phi/S} \equiv \frac{\partial(\ln v_\Phi)}{\partial(\ln v_S)}, \quad R_{\rho/S,P} \equiv \frac{\partial(\ln \rho)}{\partial(\ln v_{S,P})}, \quad (1\text{--}3)$$

where v_S is the seismic shear wave velocity, v_P is the seismic P-wave velocity, v_Φ is the seismic bulk velocity, and ρ is density, provide a way to address this question. Karato and Karki (2001) used mineral physics calculations to estimate effects of temperature and composition on these ratios and find that temperature effects (through anharmonicity and anelasticity) can explain most of the seismically inferred depth variations except for negative values of $R_{\rho/S}$ and $R_{\Phi/S}$ and $R_{S/P} > 2.7$ near the base of the mantle that require chemical effects. Chemical compositional changes in close proximity to the base of the mantle are not surprising since the D'' layer is likely a chemical as well as a thermal boundary layer (Loper and Lay, 1995). However, the depth variations of these ratios do not require chemical compositional changes in the mid-lower mantle.

Chemical compositional effects near the base of the mantle are also suggested by the free oscillation inversion studies of Ishii and Tromp (1999), who find dense, but seismically slow (and presumably hot) regions at the bottom of the mantle. On the other hand, Romanowicz (2001) reports inversions of free oscillation data for deep mantle seismic and density structure that yield the opposite correlation between density and seismic velocity and fit the data equally well. Whether or not the free oscillation data provide robust correlations between density and seismic velocity anomalies near the base of the lower mantle, any compositional changes possibly revealed by these data can reside in the D'' layer and need not be associated with a thick, isolated lower mantle region.

4. Convection style for a compositionally dense, heat source rich, deep lower mantle

The nature of convection in a mantle containing a thick, deep layer of compositionally heavy material with a large heat source concentration is illustrated by the two-dimensional model calculation of Kellogg *et al.* (1999) shown in Figure 2. A cold downwelling at the right edge of the box depresses the interface between the compositionally distinct layers and almost reaches the bottom of the box. The circulation in the upper layer is completed by a warm upwelling on the left edge of the box that originates from the interface between the layers; on the left-side of the box the interface lies almost in the middle of the box. Heavy, downwelling material in the upper layer pushes the lower layer material toward the left and piles it up on the left side of the box. The lower layer is relatively uniformly hot compared with the relatively cold upper layer mainly due to the enhanced heat source concentration in the lower layer. Viscosity is temperature dependent in the model and the cold downwelling in the upper layer is relatively viscous. The lower layer is less viscous than the upper layer

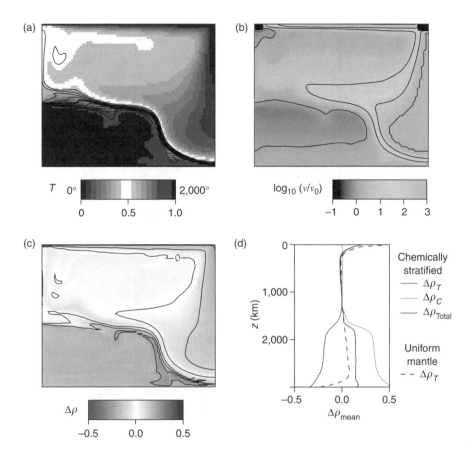

Figure 2 A two-dimensional numerical model of convection in a box containing a compositionally dense lower layer enriched in heat-producing elements. (a) Temperature (colors) and
contours of composition. A thermal boundary layer exists at the top of the lower layer;
(b) viscosity; (c) Density variations relative to the mean density at a reference depth in
the upper layer; and (d) Depth profiles of laterally averaged density variations from a reference value. (Reprinted with permission from Kellogg, L. H., Hager, B. H. and van der
Hilst, R. D., "Compositional stratification in the deep mantle," *Science* **283**, 1883–1884
(1999). Copyright 1999, American Association for the Advancement of Science.) (See
Colour Plate IV.)

(aside from the downwelling) because the lower layer is so hot. The density of the lower layer
is higher than that of the upper layer, but the overall density difference is subdued compared
with the compositional density contrast alone by the ameliorating effect of high lower layer
temperatures on layer density. Because of its cold temperature, the downwelling region of
the upper layer is essentially as dense as the lower layer explaining why the downwelling is
able to push lower layer material aside and sink almost to the bottom of the box.

Figure 3 shows possible seismic shear wave velocity variations V_S in the model of Kellogg
et al. (1999). Variations in V_S are due to both temperature and composition and the panels in
Figure 3 are based on the thermal and compositional variations in the solution of Figure 2. In
(a), (b) it is assumed that the compositional density increase in the lower layer is due to Fe (Si).

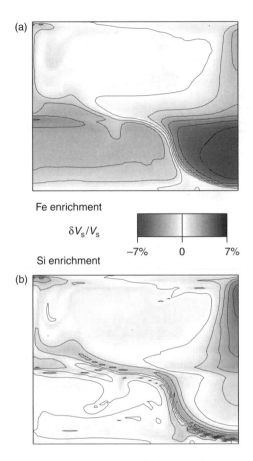

Figure 3 Seismic velocity variations in the model of Figure 2. Colors denote lateral variations in $\delta V_S/V_S$, where V_S is the seismic shear wave velocity (a,b). The compositional density increase in the lower layer is assumed to be due to Fe, Si. (Reprinted with permission from Kellogg, L. H., Hager, B. H. and van der Hilst, R. D., "Compositional stratification in the deep mantle," *Science* **283**, 1883–1884 (1999). Copyright 1999, American Association for the Advancement of Science.) (See Colour Plate V.)

The main message of the figure is that it is possible for chemistry and temperature to conspire in such a way as to offset each other's effect on seismic shear wave velocity so that there is little difference in V_S between the upper and lower layers. However, as seen in the right panel of Figure 3, even when there is little difference in V_S between the upper and lower layers, the thermal boundary layer at the interface between the layers stands out as an anomaly in V_S. This is an important point regarding the seismic detectability of a compositionally dense, heat source rich, deep lower mantle. We will return to this point later and show that if such a layer existed, it would be seismically detectable by virtue of the effect of the interface thermal boundary layer on V_S.

The model of Kellogg *et al.* (1999) has been studied in three dimensions by Tackley (2002). His results are shown in Figure 4 for a situation in which the heat source rich, compositionally heavy layer occupies the lower 30% of the wide ($4 \times 4 \times 1$) rectangular box prior to the onset

Composition   Temperature (residual)

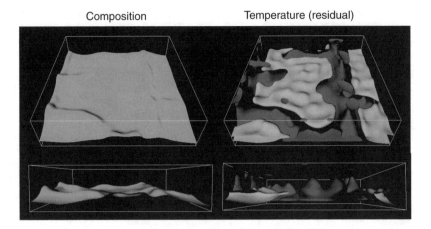

Figure 4 Isosurfaces of composition and residual temperature in a three-dimensional numerical model
of convection in a wide box containing a compositionally dense lower layer enriched in
heat-producing elements. The composition isosurface reveals the undulating character of
the top of the lower layer. The blue residual temperature isosurface reveals where upper
layer downwellings impinge on the lower layer. The red residual temperature isosurface
shows the nature of hot upwelling ridges and plumes in the upper layer. The numerical
results have been filtered to approximate seismic tomographic resolution. (Tackley, P. J.,
"The strong heterogeneity caused by deep mantle layering," *Geochemistry, Geophysics,
Geosystems* **3**(4), 10.1029/2001GC000167 (2002). Copyright 2002, American Geophysical Union. Reproduced/modified by permission of the American Geophysical Union.)
(See Colour Plate VI.)

of convection. The interface between the layers, as shown by the two views of the constant
composition surface in the left panel, has bumps and depressions arising mainly from the
ability of downwellings in the upper layer to push aside lower layer material. The residual
temperature isosurfaces shown in the right panel (surfaces of constant temperature difference
from the horizontal average) reveal both the effect of the upper layer downwellings on moving
aside lower layer material (seen in the cold isosurface) and the nature of the hot upwellings
(plumes and ridges) in the upper layer (seen in the hot isosurface).

Figure 5 shows additional views of the temperature field and a proxy of the seismic velocity
field for the model of Figure 4. The seismic velocity proxy is a linear combination of temperature T and composition C, with the relative contributions of T and C arbitrarily chosen
to show little contrast between the upper and lower layers. The particular combination of T
and C in Figure 5 thus represents the possibility that temperature and composition conspire
to mitigate the seismic velocity contrast between the upper and lower layers as discussed
above in connection with Figure 3. The leftmost panels show vertical cross-sections of T
(top) and $T - 0.65C$ (bottom). The temperature cross-section reveals how hot the lower layer
is compared with the upper layer; it also shows the interface thermal boundary layer and
hot upwellings and cold downwellings in the upper layer. The cross-section of the seismic
velocity proxy shows that while there is essentially little overall seismic velocity contrast
between the layers, the interface thermal boundary layer and the upper layer upwellings
have prominent anomalies. The six remaining panels of Figure 5 show T and $T - 0.65C$ in
horizontal cross-sections at three different depths in the box, the deepest one being almost
entirely in the lower layer, the shallowest one sampling the upper layer only, and the third

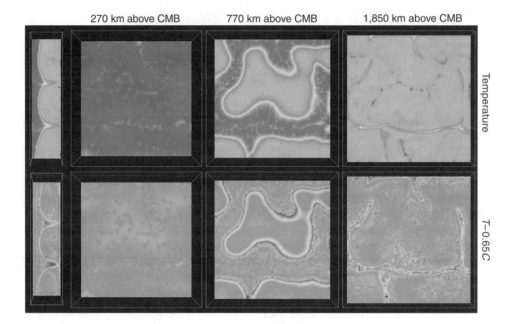

Figure 5 Views of the temperature field and a proxy of the seismic velocity field for the model
of Figure 4. The seismic velocity proxy is a linear combination of temperature T and
composition C. The leftmost panels show vertical cross-sections of T (top) and $T - 0.65C$
(bottom). The six remaining panels of the figure show T and $T - 0.65C$ in horizontal
cross-sections at three different depths in the box, the deepest one being almost entirely in
the lower layer, the shallowest one sampling the upper layer only, and the third one cutting
through both layers. The trace of the hot upwelling as it emerges from the thermal boundary
layer at the top of the lower layer stands out clearly in the field of the seismic velocity proxy.
After Tackley (2001). (Tackley, P. J., "The strong heterogeneity caused by deep mantle
layering," *Geochemistry, Geophysics, Geosystems* **3**(4), 10.1029/2001GC000167 (2002).
Copyright (2002). American Geophysical Union. Reproduced/modified by permission of
the American Geophysical Union.) (See Colour Plate VII.)

one cutting through both layers. The deep cross-sections show a relatively uniformly hot
lower layer with little variation in the seismic velocity proxy. The horizontal cross-sections
at 770 km above the bottom of the box (the CMB) cut both hot lower layer material and
cold downwellings in the upper layer; these features are associated with contrasts in the seismic velocity proxy. The horizontal cross-sections at 1,850 km above the bottom of the box
show thermal contrasts between hot upwellings and cold downwellings in the upper layer
and prominent signals in the seismic velocity proxy. The important result from this model
and from the two-dimensional model discussed above is that thermal anomalies such as the
interface thermal boundary layer show up clearly as seismic velocity anomalies even if the
effects of temperature and composition act to minimize the seismic velocity contrast between
the upper and lower layers. We will show below that the interface thermal boundary layer
would be seismically detectable if a compositionally dense, heat source rich region really
existed in the lower half of the Earth's lower mantle.

Problems with a compositionally dense, heat source rich, deep lower mantle in the real
Earth are discussed in the following subsections.

4.1. Excessive heating of the lower layer

One of the main problems with the existence of a compositionally dense, heat source rich, deep lower mantle in the real Earth is how hot the layer becomes because the large amount of heat generated in it cannot escape without being conducted through the large thermal resistance of the interface thermal boundary layer. The layer becomes so hot that it would be partially molten, in disagreement with the known structure of the Earth's lower mantle (except perhaps for isolated partially molten regions at the CMB). The high temperature of the lower layer has been shown above in the two- and three-dimensional model calculations. The high temperature of the lower layer in the three-dimensional calculation is further illustrated by the depth profile of horizontally averaged temperature shown in Figure 6. The average temperature at the base of the mantle is about a factor of 1.6 larger than it is in an adiabatic model not having the heat source rich, compositionally dense, isolated lower layer. Figure 6 also shows that when the chemically distinct lower layer occupies only 10% of the volume of the box in the motionless state, there is a relatively small increase in the average bottom boundary temperature. This is due in part to the smaller volume of the heat source rich lower layer. In addition, when the volume of the lower layer is sufficiently small, convection can gather the dense lower layer material into individual piles thereby increasing the surface area of the material and its ability to cool (Tackley, 2002). In this case the CMB is not everywhere blanketed by the compositionally dense material. It needs to be emphasized that the surface heat flux is the same in all the cases in Figure 6.

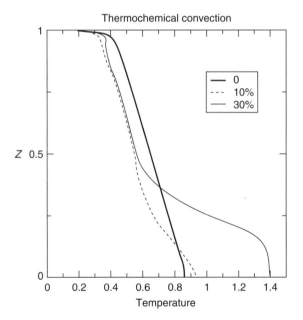

Figure 6 Depth profiles of horizontally averaged temperature in the model of Figure 4 (the compositionally dense lower layer fills 30% of the box in the undisturbed state) and in two other models (no chemically dense lower layer 0%, the compositionally dense lower layer fills 10% of the box in the undisturbed state). (Tackley, P. J., "The strong heterogeneity caused by deep mantle layering," *Geochemistry, Geophysics, Geosystems* **3**(4), 10.1029/2001GC000167 (2002). Copyright 2002, American Geophysical Union. Reproduced/modified by permission of the American Geophysical Union.)

The problem with overheating and melting of the lower layer is identical to the problem recognized by Schubert and Spohn (1981) in connection with proposals that the mantle below 660 km depth was an isolated region with a primordial concentration of radioactivity. In this classical two-layer mantle convection model the heat produced in the lower mantle must be conducted through the interface thermal boundary layer on its way to the surface. The enhanced thermal resistance of this boundary layer produces a large temperature drop across it and causes the excessively high temperatures of the lower layer. This fundamental problem with the separation of a depleted upper mantle from a primordial lower mantle persists even when the undepleted lower layer occupies only a fraction of the volume of the lower mantle. The only way this difficulty can be avoided is if the lower layer is a relatively small fraction of the lower mantle as in the 10% case of Figure 6.

4.2. Compatibility of high $^3He/^4He$ with primordial concentrations of radiogenic elements

One of the geochemical pieces of evidence cited in support of a compositionally distinct, isolated layer in the lower mantle is the high $^3He/^4He$ ratio in OIB. This is interpreted as requiring incomplete outgassing of the mantle in order to preserve a relatively high concentration of nonradiogenic ^3He. One way to achieve this end is to sequester the ^3He in an isolated reservoir within the mantle. However, if that isolated reservoir is a primordial one with enhanced concentrations of radiogenic elements, then it is not clear that the reservoir would have a high $^3He/^4He$ signal because of the increased production of radiogenic ^4He.

4.3. Lack of seismic detection of an interface thermal boundary layer

We have seen in the model calculations discussed above that even when competing effects of composition and temperature produce little net contrast in seismic shear wave velocity between the upper and lower layers, the interface thermal boundary is still a prominent feature. Vidale *et al.* (2001) have shown that the boundary layer would be seismically detectable if it really existed in the mantle; there are no seismic detections of such an interface in the lower mantle. The model of Vidale *et al.* (2001) is sketched in Figure 7. A sharp chemical boundary is assumed to exist in the mid-lower mantle at 1,770 km depth. Thermal boundary layers above and below the chemical boundary are represented as linear gradients over a depth of 200 km. A temperature increase across each boundary layer of $\Delta T = 600$ K is assumed to accommodate a large heat flux from the lower layer, although this ΔT is inadequate to transfer a large fraction of the Earth's heat loss. The assumed ΔT corresponds to a 0.1 km/s peak seismic P-wave velocity anomaly. The depth profile of seismic P-wave velocity in Figure 7 shows perturbations associated with the interface thermal boundary layers, but otherwise it passes smoothly from the upper to the lower part of the lower mantle in accord with the nearly adiabatic character of the lower mantle as seen in Earth models like PREM (Dziewonski and Anderson, 1981). The model assumes perfect cancellation of thermal and compositional variations outside the thermal boundary layers; seismic detectability of the compositionally distinct lower layer therefore relies solely on the thermal boundary layers at the top of the layer. The cancellation of thermal and compositional effects on seismic velocity comes about because the seismic velocity decrease due to the temperature excess of the lower layer is offset by the presumed seismic velocity increase in the compositionally distinct material of the lower layer. This conservative model assumption is adopted even though the greater density of the lower layer would normally be expected to decrease, rather than increase, the

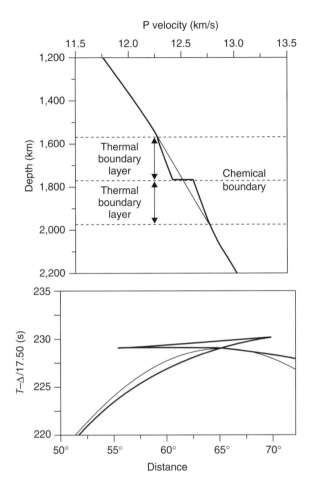

Figure 7 (Top) Seismic P-wave velocity structure in the model of Vidale *et al.* (2001) having a mid-mantle chemical boundary with thermal boundary boundary layers on either side (thick line). The thin line shows the depth profile of seismic P-wave velocity in the PREM model. The two seismic P-wave distributions are identical except for the anomalous thermal boundary layer regions. (Bottom) Seismic P-wave travel time versus distance in the model of Vidale *et al.* (2001) (thick lines) and in the PREM model (thin line). Seismic P-wave velocity anomalies in the thermal boundary layers produce a triplication in the travel time versus distance curve. (Vidale, J. E., Schubert, G. and Earle, P. S., "Unsuccessful initial search for a mid-mantle chemical boundary with seismic arrays," *Geophysical Research Letters* **28**, 859–862 (2001). Copyright 2001, American Geophysical Union. Reproduced/modified by permission of American Geophysical Union.)

seismic velocity. The P-wave seismic velocity is smaller (larger) than that of the surroundings (PREM) in the thermal boundary layer just above (below) the chemical interface since the material of the boundary layer is identical in composition to the upper (lower) layer but it is at a higher (lower) temperature.

Figure 7 also shows the P-wave travel time (computed by ray tracing) as a function of distance for the depth profile of seismic velocity shown in the figure. A distinct triplication

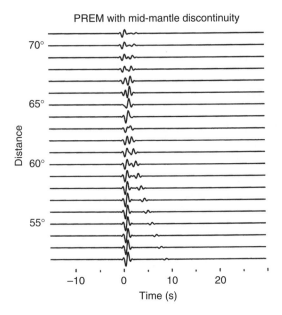

Figure 8 Synthetic seismograms from the model of Vidale *et al.* (2001) with a mid-mantle chemical boundary. The triplication in Figure 7 produces waveform complexity in the distance range from about 58° to 68°. (Vidale, J. E., Schubert, G. and Earle, P. S., "Unsuccessful initial search for a mid-mantle chemical boundary with seismic arrays," *Geophysical Research Letters* **28**, 859–862 (2001). Copyright 2001, American Geophysical Union. Reproduced/modified by permission of American Geophysical Union.)

occurs for short-period P-waves. Synthetic seismograms that would result from the P-wave seismic velocity structure of Figure 7 and the assumption of a flat horizontal chemical interface are depicted in Figure 8 from Vidale *et al.* (2001). The seismograms show multiple arrivals separated by up to 2 s in the distance range from about 58° to 68°. Local topography of a chemical interface would influence the range and strength of the arrivals. Vidale *et al.* (2001) carried out a search for multiple arrivals in the distance range of about 60–70° in seismograms from 10 earthquake events using data from the Northern and Southern California Seismic Networks. No evidence of waveform variations similar to those in Figure 8 was found. While the study of Vidale *et al.* (2001) is geographically limited, there are no reports in the literature of waveform variations that would be diagnostic of a lower mantle chemical interface from any other region. These authors concluded that the absence of the diagnostic waveform signature in observations favors models in which the mid-mantle is well mixed. A similar conclusion has been reached by Castle and van der Hilst (2003) who searched unsuccessfully for the transmitted S to P conversions that would be expected from a mid-mantle chemical interface. Chemical boundaries in the lower mantle cannot be hidden from seismic detection and the seismic evidence to date shows that such boundaries do not exist.

5. Summary and conclusions

Seismic tomography reveals that mantle convection occurs in a single layer extending from the surface to a depth of at least the middle of the lower mantle. At least some convective elements (slabs) from the upper mantle penetrate essentially to the CMB. It is no longer tenable to consider the base of the transition zone at 660 km depth as the bottom of the upper layer of a two-layer mantle convection system even though seismic tomography shows that some slabs are apparently impeded at this depth as they descend into the mantle. It is difficult to reconcile this picture with geochemical data that are interpreted to require compositionally distinct, isolated reservoirs in the mantle. One proposal envisions the bottom half of the lower mantle as a primordial, compositionally distinct, isolated layer. This layer provides a convenient storage site for radioactive elements that it is argued cannot be in the upper half of the mantle. The excess chemical density of the layer would make it heavy enough to remain at the bottom of the mantle even though its internal heat sources would heat it and make it thermally buoyant. Although descending slabs might be able to push the lower layer material aside to some extent, the chemically heavy lower layer would tend to stop slabs from sinking deep into the lower mantle and help explain the disappearance of the slab signal from seismic tomographic images of the bottom half of the mantle.

There are, however, a number of problems with this model, the discussion of which has been the main focus of this paper. In addition, we have shown that the heat flow and seismological observations motivating the model can be understood in other ways. With regards to the Earth's heat flow, there are fundamental uncertainties in how much heat is actually escaping from the Earth's interior and what fraction of it is to be attributed to mantle radioactivity versus secular cooling of the Earth. Basically, we do not know the Urey number of the Earth. The inference of how much radioactivity there is in the Earth's mantle from measured concentrations of radioactive elements in MORB could account for very little of the Earth's heat loss, but the mantle heat source concentration inferred this way is highly uncertain. Cosmochemical estimates of the amount of radioactivity in the Earth allow for a larger fraction of the Earth's heat loss to derive from mantle radioactivity, but this approach is limited by our knowledge of which type of chondritic meteorite really represents the building blocks of the Earth. Quite reasonable models of the amount of radioactivity in the mantle are consistent with the heat flow from the Earth without the necessity of hypothesizing the existence of a mantle region enriched in radioactive heat sources.

With regards to the disappearance of the slab signal in seismic tomographic images of the lower mantle, there are many possible explanations that have nothing to do with a compositionally dense layer in the lower mantle. It may simply be that the seismic signal is weak even though the slabs are there. It may also be that descending slabs are reabsorbed into the general mantle circulation as they approach the bottom of the mantle as happens in numerical models of convection. The depth to which a slab sinks in the mantle or the rate at which it is reabsorbed into the surroundings depends on its buoyancy, temperature, and rheology; the apparent disintegration of slabs in the lower mantle may have more to do with their temperature and rheology than with their ponding on top of a chemically dense lower layer.

A major difficulty with a heat source rich, isolated lower mantle layer is the difficulty of getting its heat out. The radioactively generated heat in the layer needs to be conducted across thermal boundary layers at the top of the lower layer and the bottom of upper layer. This relatively inefficient mode of heat transfer raises the temperatures in the lower layer so high that it should be substantially partially molten, in conflict with seismic evidence for the solidity of the lower mantle. This problem has been recognized for decades in connection with the proposal that the entire lower mantle below 660 km depth is an isolated region with primordial concentrations of radioactive elements. The problem does not go away simply by shrinking the size of the lower layer, especially if the same amount of overall heat production is put into a smaller volume.

The most serious problem with the existence of a chemically distinct, isolated layer at the bottom of the lower mantle is the lack of seismic evidence for it. We have emphasized that such a layer would be visible to seismic probing because of the thermal boundary layer at its top. While it is possible to imagine that effects of temperature and composition could offset each other in such a way as to give a smoothly varying depth profile of seismic velocity consistent with an apparently adiabatic lower mantle, the perturbation of the seismic velocity by the interface thermal boundary layer could not be suppressed this way; it would provide a characteristic signal in seismic waveforms that is not observed.

The reconciliation of some geochemical observations with a model of whole mantle convection does not lie in an isolated primordial layer buried deep in the lower mantle any more than it did when the entire lower mantle was thought by some to be the isolated reservoir.

6. Acknowledgments

Thanks are due to Louise Kellogg, Paul Tackley, and John Vidale for making available electronic versions of their figures. Supported by the Planetary Geology and Geophysics Program of NASA under grant NAG 5-3863 and by a grant from the Institute of Geophysics and Planetary Physics at the Los Alamos National Laboratory.

References

Albarede, F. and van der Hilst, R. D., "New mantle convection model may reconcile conflicting evidence," *EOS Trans. AGU* **80**(45), 535–539 (1999).

Allegre, C. J., Staudacher, T., Sarda, P. and Kurz, M., "Constraints on evolution of Earth mantle from rare gas systematics," *Nature* **303**(5920), 762–766 (1983).

Allegre, C. J., Hofmann, A. and O'Nions, K., "The argon constraints on mantle structure," *Geophys. Res. Lett.* **23**(24), 3555–3557 (1996).

Bercovici, D., Schubert, G. and Glatzmaier, G. A., "Three-dimensional spherical models of convection in the Earth's mantle," *Science* **244**(4907), 950–955 (1989a).

Bercovici, D., Schubert, G., Glatzmaier, G. A. and Zebib, A., "Three-dimensional thermal convection in a spherical shell," *J. Fluid Mech.* **206**, 75–104 (1989b).

Calderwood, A. R., "The distribution of U, Th, and K in the Earth's crust, lithosphere, mantle and core: Constraints from an elemental mass balance model and the present day heat flux," In: *SEDI 2000*, the 7th Symposium on Studies of Earth's Deep Interior, Exeter, UK, Abstract Vol. S9.4 (2000).

Castle, J. C. and van der Hilst, R. D., "Searching for seismic scattering off mantle interfaces between 800 and 2000 km depth," *J. Geophys. Res.* **108**, 2095, doi: 10.1029/2001JB000286 (2003).

Davies, G. F., "Thermal histories of convective Earth models and constraints on radiogenic heat production in the Earth," *J. Geophys. Res.* **85** (B5), 2517–2530 (1980).

Davies, G. F., "Ocean bathymetry and mantle convection, 1. Large-scale flow and hot spots," *J. Geophys. Res.* **93**, 10467–10480 (1988).

Dziewonski, A. M. and Anderson, D. L., "Preliminary reference Earth model," *Phys. Earth Planet. Inter.* **25**(4), 297–356 (1981).

Fukao, Y., Obayashi, M., Inoue, H. and Nenbai, M., "Subducting slabs stagnant in the mantle transition zone," *J. Geophys. Res.* **97**, 4809–4822 (1992).

Grand, S. P., van der Hilst, R. D. and Widiyantoro, S., "Global seismic tomography: A snapshot of convection in the Earth," *GSA Today* **7**(4), 1–7 (1997).

Hofmann, A. W., "Chemical differentiation of the Earth: The relation between the mantle, continental crust, and oceanic crust," *Earth Planet. Sci. Lett.* **90**, 297–314 (1988).

Hofmann, A. W., "Mantle geochemistry: The message from oceanic volcanism," *Nature* **385**, 219–229 (1997).

Ishii, M. and Tromp, J., "Normal-mode and free-air gravity constraints on lateral variations in velocity and density of Earth's mantle," *Science* **285**(5431), 1231–1235 (1999).

Jacobsen, S. B. and Wasserburg, G. J., "The mean age of mantle and crustal reservoirs," *J. Geophys. Res.* **84**(B13), 7411–7427 (1979).

Karato, S. and Karki, B. B., "Origin of lateral variation of seismic wave velocities and density in the deep mantle," *J. Geophys. Res.* **106**(B10), 21771–21783 (2001).

Kellogg, L. H., Hager, B. H. and van der Hilst, R. D., "Compositional stratification in the deep mantle," *Science* **283**(5409), 1881–1884 (1999).

Loper, D. E. and Lay, T., "The core–mantle boundary region," *J. Geophys. Res.* **100**(B4), 6397–6420 (1995).

McDonough and Sun, "The composition of the Earth," *Chem. Geol.* **120**, 223 (1995).

Malamud, B. D. and Turcotte, D. L., "How many plumes are there?," *Earth Planet. Sci. Lett.* **174**(1–2), 113–124 (1999).

O'Nions, R. K., Evensen, N. M. and Hamilton, P. J., "Geochemical modeling of mantle differentiation and crustal growth," *J. Geophys. Res.* **84**(B11), 6091–6101 (1979).

Pollack, H. N., Hurter, S. J. and Johnson, J. R., "Heat flow from the Earth's interior: Analysis of the global data set," *Rev. Geophys.* **31**(3), 267–280 (1993).

Romanowicz, B., "Can we resolve 3D density heterogeneity in the lower mantle?," *Geophys. Res. Lett.* **28**, 1107–1110 (2001).

Schubert, G. and Spohn, T., "Two-layer mantle convection and the depletion of radioactive elements in the lower mantle," *Geophys. Res. Lett.* **8**(9), 951–954 (1981).

Schubert, G., Stevenson, D. and Cassen, P., "Whole planet cooling and the radiogenic heat source contents of the Earth and Moon," *J. Geophys. Res.* **85**(B5), 2531–2538 (1980).

Schubert, G., Turcotte, D. L. and Olson, P., *Mantle Convection in the Earth and Planets.* Cambridge University Press, Cambridge (2001).

Sharpe, H. N. and Peltier, W. R., "A thermal history model for the Earth with parameterized convection," *Geophys. J. R. Astr. Soc.* **59**(1), 171–203 (1979).

Sleep, N. H., "Hotspots and mantle plumes: some phenomenology," *J. Geophys. Res.* **95**(B5), 6715–6736 (1990).

Stacey, F. D., "The cooling Earth: a reappraisal," *Phys. Earth Planet. Inter.* **22**(2), 89–96 (1980).

Stevenson, D. J. and Turner, J. S., "Fluid models of mantle convection," In: *The Earth: Its Origin, Structure and Evolution* (Ed. M. W. McElhinny) pp. 227–263. Academic Press, London (1979).

Tackley, P. J., "Strong heterogeneity caused by deep mantle layering," *Geochem. Geophys. Geosyst.* **3**(4), 1024, doi: 10.1029/2001GC000167 (2002).

Turcotte, D. L., "On the thermal evolution of the Earth," *Earth Planet. Sci. Lett.* **48**(1), 53–58 (1980).

Turcotte, D. L. and Schubert, G., *Geodynamics Applications of Continuum Physics to Geological Problems.* Wiley, New York (1982).

Turcotte, D. L. and Schubert, G., "Tectonic implications of radiogenic noble gases in planetary atmospheres," *Icarus* **74**(1), 36–46 (1988).

van der Hilst, R. D. and Karason, H., "Compositional heterogeneity in the bottom 1000 kilometers of Earth's mantle: Toward a hybrid convection model," *Science* **283**(5409), 1885–1888 (1999).

van der Hilst, R. D., Widlyantoro, S. and Engdahl, E. R., "Evidence for deep mantle circulation from global tomography," *Nature* **386**(6625), 578–584 (1997).

Vidale, J. E., Schubert, G. and Earle, P. S., "Unsuccessful initial search for a midmantle chemical boundary with seismic arrays," *Geophys. Res. Lett.* **28**, 859–862 (2001).

Wasserburg, G. J. and Depaolo, D. J., "Models of Earth structure inferred from neodymium and strontium isotopic abundances," Proceedings of the National Academy of Sciences of the United States of America, **76**(8), 3594–3598 (1979).

Wasson, J. T. and Kallemeyn, G. W., "Compositions of chondrites," *Phil. Trans. R. Soc. Lond.* A **325**, 535–544 (1988).

3 Electromagnetic and topographic coupling, and LOD variations

Dominique Jault

LGIT, Université Joseph-Fourier de Grenoble, BP 53, 38041 Grenoble Cedex 9, France

The role played by different core processes in the changes in the Earth's rotation is assessed and fully dynamical models of the torsional Alfvén waves inside the fluid core are reviewed. These waves, first studied by Braginsky (1970), consist of geostrophic circulation. They have decadal periods and yield time changes in core angular momentum. They arise from small departures from an hypothetical quasi-static state, where the total action of the Lorentz force on the geostrophic cylinders cancels out. They cause torques acting on the mantle. Simple models of the torsional waves that rely only on zonal averages of the magnetic field have incorporated electromagnetic coupling to the mantle. They, however, need some correction. In addition, only a kinematic approach of the topographic coupling, caused by a nonaxial symmetry of the fluid cavity, has been successfully attempted to date. Taking into account uncertainties in the height of the core–mantle topography and in the electrical conductivity of the deep mantle, it turns out that, in the present state of core modelling, the pressure, gravity and electromagnetic torques acting on the mantle may all produce decade changes in the length of the day with a magnitude comparable to the observations.

1. Introduction

Our understanding of the part played by the core in the changes in the Earth's rotation has been much improved during the recent years because of longer and more accurate geodetic series, refined modelling of the atmospheric and oceanic contributions to the Earth's angular momentum budget and progress in dynamo theory. We have benefited also from a wealth of detailed studies about core–mantle coupling at different time scales. The areas of Earth's rotation studies where processes taking place in the Earth's core or at the core–mantle boundary (CMB) are important are now well delimited. As far as the Earth's spin rate is concerned, only decade and perhaps part of the longer period variations are caused by such processes. Motions in the Earth's core are probably inoperative in the excitation of the polar motion but a resonance with the quasi-diurnal inertial core mode of nearly rigid rotation about an equatorial axis plays an important part in the response of the Earth (nutations of its rotation axis) to gravitational torques from the moon and the sun. This review deals mainly with the origin of the decade variations in the length of day (LOD) and the nutation problem is mentioned more briefly. The first question is indeed much more intricate and touches the theories of the geomagnetic secular variation and the geodynamo. However, the two problems should not be lightly dissociated since modelling of forced nutations may eventually constrain parameters (electrical conductivity of the lower mantle, energy of the small-scale magnetic field at the core surface, CMB topography) that are crucial in the modelling of the decade variations in the Earth's spin rate. I review now other geophysical informations on these parameters.

Models of CMB topography have been inferred from shear and compressional velocities within the mantle: seismic and anomalies are converted into density anomalies, which drive mantle convection and cause a dynamical topography of both the Earth's surface and the CMB. The calculated height above a reference ellipsoid of the latter surface is of the order of a few kilometres (Forte and Peltier, 1991; Defraigne *et al.*, 1996). Equipotential surfaces of the gravity field are also obtained. At the Earth's surface, the observed geoid constrains the modelling, whereas Forte and Peltier (1991) (and respectively Defraigne *et al.*, 1996) inferred that the deviation from an ellipsoid of the gravity equipotential surface is of the order of 500 (150) m at the CMB and of the order of 100 (30) m at the inner core boundary. The electrical conductivity of most of the lower mantle is rather well known. Studies of electrical currents induced in the mantle by external fluctuations (magnetotelluric and magnetic observatory data) show that the conductivity is of the order of $1\,S\,m^{-1}$ at the top of the lower mantle (Schultz *et al.*, 1993; Petersons and Constable, 1996) while high-pressure experiments show that it is of the order of $1–10\,S\,m^{-1}$ in most of the lower mantle. The remaining incertainties stem from the dependence of the electrical conductivity on the temperature and the aluminium content (Xu *et al.*, 1998). The combination of experiments with geophysical studies may yet narrow the range of permissible values (Dobson and Brodholdt, 2000). We shall see in Section 5 that the electrical currents circulating in the lower mantle with conductivity $5\,S\,m^{-1}$ are too weak to couple efficiently core and mantle on the decade time scale. Only the electrical currents at the bottom of the mantle may be intense enough to participate to core–mantle coupling, assuming a high conducting layer to be present there.

A variety of seismological investigations have been recently focused on the mantle region just above the CMB (see the collected articles in Gurnis *et al.*, 1998). The D'' layer encompassing 200–300 km of the lowermost mantle has long ago been identified as a region of low seismic velocity gradients. It shows intense lateral variations, in particular of thickness, and it includes regions where seismic waves are anisotropic. In places, it is separated from the normal mantle by a velocity discontinuity (or a high velocity layer). An ultralow velocity zone (ULVZ) has recently been discovered (see the review of Garnero, 2000) at the bottom of D''. Its thickness is 5–50 km where it has been detected (a third of the probed areas). The low velocity in this region may be the result of partial melting. This patchy zone may be much more dense than the surrounding mantle causing CMB topography. Only a week perturbation of the gravity potential would be associated with such a topography. Increase of the electrical conductivity in the ULVZ is likely if it is partially molten. Its magnitude would depend on the chemical composition of the zone. These seismic observations may be taken as an indication of lateral variation of the electrical conductivity at the bottom of the mantle. Anisotropy of the conductivity is also a possibility.

Recently, seismologists have thoroughly investigated a possible differential rotation between the inner core and the solid mantle. Its determination would obviously yield an invaluable constraint on models of the Earth's rotation. Differential travel times of seismic waves and free oscillations have been examined, whereas Vidale *et al.* (2000) have used temporal changes in the scattering of seismic waves inside the inner core to suggest a differential rotation between the inner core and the mantle at $0.15°$ per year between 1971 and 1974 (see also the references inside this article). This method has the potential to monitor the inner core axial rotation for periods of a few years. Meanwhile, we do not know whether the detected rotation is steady or participates in the decadal changes in the Earth's axial rotation reflected in LOD fluctuations.

Magnetic field observations remain the principal source of information on core dynamics. In this chapter, I keep the usual terminology and I refer to the time series of the Earth's magnetic field as secular variation (SV) data. We are now expecting a dramatic improvement in the accuracy of SV data after a few years of continuous satellite recordings of the three components of the magnetic field, but, despite all the efforts of data analysis, we still have to rely heavily on dynamo models to unravel the mysteries of core dynamics. Rapid changes of the geostrophic velocity appear intertwined with a slow evolution of an important force equilibrium inside the core (Taylor, 1963). They entail changes in core angular momentum, which attest, conversely, the inner working of the Earth's dynamo. The time scales of the different mechanisms, inside the core, are measured against the magnetic diffusive time τ_d, of the order of a few tens of thousands of years. In the presence of rapid rotation and of a strong ambient magnetic field, the long lengthscale waves riding inside the core (MC waves; Fearn et al., 1988) have periods τ_{MC} of the order of τ_d/Λ, where the Elsasser number Λ gives the strength ratio of the magnetic force to the rotation force, on time scales comparable to τ_d and longer. Because Λ is probably of order unity, the general opinion is that these MC waves do not represent the observed rapid variations of the Earth's magnetic field. Thus, the Alfvén torsional waves, which involve only the geostrophic part of the motion, stand out because of their rapid periods (tens of years). They were first described 30 years ago by Braginsky (1970) in a spherical cavity. This review discusses how this theory has grown with the contact of the geophysical data that have since been collected. I rely also on recent investigations of the convective dynamo (Bell and Soward, 1996; Bassom and Soward, 1996) to suggest a possible extension to the case of bumpy CMB and/or inner core surface. Finally, I remark that Alfvén torsional waves are not easily excited within the parameter range where fully consistent numerical models of the geodynamo currently operate. In these models, the ratio between the magnetic diffusivity η and the kinematic viscosity v is decreased to enable dynamo action. As a result, the spin-up time scale τ_E becomes shorter than the period τ_{TA} (see Eq. (3.29)) of the torsional waves. Introducing the magnetic Prandtl number $P_m = v/\eta$, we obtain $\tau_{TA}/\tau_E \simeq P_m^{1/2}\Lambda^{-1/2}$. With $P_m \geq 1$ and $\Lambda = O(1)$, viscous dissipation precludes propagation of the torsional waves. This situation contrasts with the actual geophysical case, where $P_m = O(10^{-6})$.

In the following section, I discuss how the core responds to torques of external origin. Next, I present the theory of torsional Alfvén waves, which gives an appealing explanation for the decade changes in the length of the day. A fourth and short section is devoted to extensions of this theory in presence of an inner core and of regions where geostrophic contours do not exist. However, simplicity goes only so far. From models of the secular variation of the Earth's magnetic field, it transpires that other core surface motions, besides Alfvén torsional waves, have also short time scales. We have to rely on kinematic theories, reviewed in Section 5, to evaluate their possible influence on core–mantle coupling.

2. Response of the core to changes in the rotation of its container

I begin with a summary of the changes in the Earth's rate of rotation and orientation in order to get the role of the core into perspective. It can be kept brief because very useful review articles, giving a lot of references, are already available (Dickey and Hide, 1991; Eubanks, 1993). There is strong evidence that exchanges of angular momentum between core and mantle occur on the decade time scale (Jault et al., 1988; Jackson et al., 1993; Jackson, 1997). These yield variations in the length of the day of up to a few milliseconds (ms). On the other hand, the role of the atmosphere in rapid changes in the rotation rate of the solid Earth is well ascertained (Rosen et al., 1990 and references therein) up to periods of a few years. The oceans and the atmosphere may play also some role on longer periods.

As an example, Abarca del Rio (1999) argued that thermal expansion of the oceans may have caused an increase in the LOD as large as 0.25 ms from 1950. Yet, the contributions of the outer fluid envelopes to the Earth's angular momentum budget, on the decade time scale, are minor. It is likely that the core does not play much role at shorter periods but we lack quantitative studies. In order to shed light on this question, it is possible either to look for discrepancies between changes in the combined angular momentum of the atmosphere and the oceans and changes in the angular momentum of the solid Earth on periods of a few years (see, e.g. Abarca del Rio *et al.*, 2000) or to study the response of the core to forcing by changes in the rotation rate of the mantle (spin-up, spin-down).

Zonal tides produce changes in the Earth's axial moment of inertia and in the angular momentum of oceanic currents. Variations in the spin-rate of the mantle ensue. From an analysis of the tidally induced changes in the LOD, Dickman and Nam (1998) concluded that the core is fully decoupled from the mantle at 9 days period, while they found that no firm conclusions are possible from the study of longer period tidal effects in the Earth's rotation. Up to the annual period, once the atmospheric contribution is removed and assuming core–mantle decoupling, the residual LOD is only a few percents of the total LOD signal. Thus, full coupling of the core with the mantle at these frequencies would be measurable since the moment of inertia of the core is 12% of that of the mantle. Then, Zatman and Bloxham (1997a) remarked that the characteristic time scale of the coupling processes between the core and the mantle can be inferred, if it is short enough, from a study of the phase difference between LOD and atmospheric angular momentum. Such a phase difference has long been sought as a clue for oceanic influences. The first studies (Eubanks *et al.*, 1985; Rosen *et al.*, 1990) found no difference between the two series. With more accurate geodetic data, the coherence between the two series has been increased to the level where oceanic contributions are now significant. Taking advantage of the improved modelling of oceanic dynamics, Dickey *et al.* (2000) have been able to combine the angular momentum of the atmosphere and of the oceans and to compare the resulting series with LOD. They have found no phase difference. It would imply that, up to the annual period at least, changes in the rate of rotation of the mantle have no significant effects on the core rotation. Finally, Dickman (2001) has just advocated incorporating, in models of LOD changes at annual and semi-annual periods, the variations in the Earth's gravity inferred from satellite laser ranging observations. He thus expects to measure accurately the extent of core–mantle coupling at annual period. In the meantime, I tentatively conclude from this discussion that the characteristic times of the coupling mechanisms between core and mantle are of the order of 1 year at least.

Accurate VLBI measurements of the nutations of the Earth's axis of rotation may give also constraints on core–mantle coupling. The VLBI series is now long enough to determine the parameters of the 18.6-year nutation. If the core were inviscid, its outer boundary spherical and the lower mantle electrically insulating, forced nutations of the Earth would yield a differential rotation between the core and its container about an equatorial axis fixed in an inertial frame. This diurnal mode, in a frame attached to the mantle, is actually the simplest possible free mode of a spherical fluid body: the 'tilt-over' mode. Since the CMB is ellipsoidal, this free mode is coupled to the mantle, it is a normal mode of the whole Earth and it entails a rotational motion of the solid Earth, the free core nutation, also observed in VLBI series. The period of this free mode (retrograde in the mantle frame) depends on the core ellipticity. In the mantle frame, the mode induces resonance in the diurnal tides. In an inertial frame, its period is close to 1 year and the forced retrograde annual nutation of the Earth is significantly modified because of the presence of the fluid core. Thus, the measurement of the amplitude of this forced nutation, together with analysis of tidal gravity data, enables to determine accurately the period of this free core mode of rotation about an equatorial axis, which, in turn, constrains the oblateness of the CMB (Gwinn *et al.*, 1986; Neuberg *et al.*, 1987). In the same way, another free mode (prograde) consists mainly of the rotation of the solid inner core about an equatorial axis. It has a longer period in an inertial frame and influences the 18.6 years nutation. Finally, the motive for this discussion is that observations of forced nutations out-of-phase with luni-solar forcing require, according to Buffett (1992), very efficient electromagnetic coupling at the diurnal frequency. In particular, the out-of-phase component of the annual retrograde nutation is very sensitive to core–mantle coupling and cannot be explained by mantle anelasticity or ocean tide loading (Mathews *et al.*, 2001). It constrains the conductivity of a thin layer above the CMB of thickness the penetration depth of diurnal signals from the core. Buffett *et al.* (2000) have just reported the result of an inversion of the most up-to-date nutation observations. The model includes a solid layer of core conductivity attached to the mantle and an energetic small scale magnetic field, such that the radial field has an uniform rms strength of 7.1×10^{-4} T over the CMB. The remaining residuals after the inversion are small but some of them may still be significant (the out-of-phase component of the prograde 18.6 years nutation). The hypothesis of high energy in the small-scale part of the magnetic field will also soon be assessed with satellite observations of the geomagnetic field (Neubert *et al.*, 2001). Finally, the model is compatible with the long spin-up time for the core advocated in the previous paragraph and with short diffusion times of electromagnetic signals through the mantle, yet the two hypotheses of highly conducting solid layer and strong small-scale magnetic field at the core surface are far from trivial. We must allow for other possibilities. In addition to the nearly rigid rotation about an equatorial axis, other inertial modes of the fluid outer core are coupled to the nutations of the solid Earth when there is topography at the CMB (Wu and Wahr, 1997). The modes with nearly diurnal periods are particularly significant. The difficulty, here, is to avoid introducing too many parameters for too few data. I conclude now this summary of the constraints that the externally driven changes in the Earth's rotation give on core–mantle coupling with a discussion of another free mode of the Earth, the Chandler wobble.

First, there is a sharp contrast with the free core nutation problem. Dissipative processes at the CMB, investigated as a damping mechanism of the Chandler wobble, can be neglected compared to mantle anelasticity (Smith and Dahlen, 1981). On the other hand, a possible role of the core in the excitation of this wobble of the Earth's rotation axis has long been debated. The Chandler excitation power shows indeed dramatic decadal changes, which have been tentatively associated with core processes such as impulses in the secular variation of the Earth's magnetic field (Gibert *et al.*, 1998). There is however growing evidence from improved modelling of oceanic circulation that the Chandler wobble is mostly excited by a combination of atmospheric and oceanic processes. Celaya *et al.* (1999) relied on a statistical analysis of a full climate model coupling the oceans and atmosphere to infer that the climate excitation of the Chandler wobble has the right amplitude and time scale. They noticed indeed that the excitation of the wobble is consistent with a stationary Gaussian process. In a complementary study, Gross (2000) used a global oceanic circulation model, constrained by actual observations, to argue that fluctuations in the pressure at the bottom of the oceans, driven by surface winds, have been the main excitation process during 1985–1996. He assumed the same value of the quality factor ($Q = 179$) of the Chandler wobble as Celaya *et al.* (1999), which is still uncertain. On longer time scales, the influence of core dynamics is plausible. However, estimates of the pressure torque acting from the core seem too small (Hide *et al.*, 1996; Hulot *et al.*, 1996) to explain past motions, with decadal periods, of the pole. In addition, the most recent and accurate data do not show decade variations (Mc Carthy and Luzum, 1996).

3. Modelling the Alfvén torsional waves inside the Earth's core

Alfvén torsional waves consist of geostrophic motions, which carry axial angular momentum. They have periods adequate to participate in the LOD variations with decade time scales and they occur naturally as the configuration of the magnetic field slowly evolves. Keeping only the necessary ingredients of a convective dynamo model (see e.g. equation (3.36) of Fearn, 1998), the momentum equations for slow- and large-scale motions are

$$2\rho(\mathbf{\Omega} \times \mathbf{u}) = -\nabla p + \mathbf{j} \times \mathbf{B} - \alpha\rho\Theta\mathbf{g}, \tag{3.1}$$

$$\nabla \cdot \mathbf{u} = 0, \tag{3.2}$$

$$\mathbf{u} \cdot \mathbf{n}|_{\Sigma} = 0, \tag{3.3}$$

where $\mathbf{g}$ is the gravity acceleration, Θ is the temperature, α is the coefficient of thermal expansion, $\mathbf{B}$ is the magnetic field, $\mathbf{j}$ is the electrical current density, p is the pressure, $\mathbf{u}$ is the velocity, $\mathbf{\Omega}$ is the spin-rate of the mantle, ρ is core density, and $\mathbf{n}$ is the outward normal to the boundary Σ of the fluid volume. The first equation represents the magnetostrophic balance between the rotation and Lorentz forces. Many terms are neglected and we shall see that it is not always consistent. Knowing $\mathbf{B}$ and Θ, $\mathbf{u}$ is determined up to an arbitrary geostrophic motion $\mathbf{u}_g$, obeying the balance

$$2\rho(\mathbf{\Omega} \times \mathbf{u}_g) = -\nabla p_g, \quad \mathbf{u}_g \cdot \mathbf{n}|_{\Sigma} = 0. \tag{3.4}$$

Geostrophic motions are independent of the coordinate z in the direction of the rotation axis. They are thus entirely defined by their streamlines on Σ, the pair of geostrophic contours Γ. Denote, respectively, z_T and z_B the z-coordinates along each upper and lower geostrophic contours. The length $H = z_T - z_B$ is an invariant of each pair of contours. The pressure p_g is constant on each cylinder $\mathcal{C}$, parallel to the rotation axis, generated by geostrophic contours. These cylinders are defined in an unique way by their total height H. There may be regions of the core where no geostrophic contours exist. I defer their discussion to Section 4. The Taylor's constraint on the right-hand side of Eq. (3.1) is a corollary of the nonuniqueness of $\mathbf{u}$:

$$\forall \mathbf{u}_g, \quad \int_{V_\Gamma} \mathbf{u}_g \cdot (\mathbf{j} \times \mathbf{B} - \alpha\rho\Theta\mathbf{g}) \, dV = 0, \tag{3.5}$$

where V_Γ denotes the region where geostrophic contours exist. By investigating the volume comprised between the cylinders $\mathcal{C}(H)$ and $\mathcal{C}(H+dH)$, it can be checked that the generalized version of the Taylor's condition given by Bassom and Soward (1996) (their equation (1.3) is equivalent to Eq. (3.5)). In the special case of a spherical cavity enclosed between two spheres of radius, respectively, b and a, the geostrophic contours are circular, the geostrophic velocity is constant on each cylinder $\mathcal{C}$, and

$$z_B = -z_T, \quad \mathbf{u}_g = u_g(s)\mathbf{e}_\phi \quad \text{at } b \leq s, \tag{3.6}$$

$$\mathbf{u}_g = u_g^{\pm}(s)\mathbf{e}_\phi \quad \text{at } s \leq b \text{ and } \pm z \geq 0, \tag{3.7}$$

where s is the distance to the rotation axis and $\mathbf{e}_\phi$ the unit azimuthal vector. Let us scale the different surface contributions to Eq. (3.5) with respect to the JB Taylor volume integral derived for an insulating mantle and a spherical core, radius a,

$$\forall \mathbf{u_g}, \quad \int \mathbf{u_g} \cdot (\mathbf{j} \times \mathbf{B})\, dV = 0, \tag{3.8}$$

with $\mathbf{j} = \mathbf{0}$ in the solid mantle: I estimate the relative error caused, on the long geodynamo time scale, by the substitution of Eq. (3.8) to Eq. (3.5). Assuming that buoyancy and electromagnetic forces are similar in strength, topographical effects, which arise because geostrophic contours deviate from perfect circles, scale as

$$\delta = \frac{h}{a} = 3 \times 10^{-4}, \tag{3.9}$$

where the height h of the CMB topography yields also the characteristic distortion of the cylinders. The importance of the Lorentz force acting in the lower mantle compared to its counterpart inside the core is measured by

$$\frac{\sigma_m \Delta}{\sigma_c a} = 3 \times 10^{-5}, \tag{3.10}$$

where Δ is the thickness of the layer of conductivity σ_m at the bottom of the mantle and σ_c is core conductivity. The viscous drag, scaled by the square root of the Ekman number E is also negligible:

$$E^{1/2} = 10^{-7}, \quad E = \frac{\nu}{\Omega a^2}. \tag{3.11}$$

Core–mantle coupling thus appears unimportant for these long time scale dynamics.

When condition (3.5) is not fulfilled, Eq. (3.1) has no solutions and it has to be modified by including at least another term, such as inertia or the viscous force (see the complete discussion by Roberts and Soward (1972)). I shall leave out a possible turbulent friction force at the solid boundaries (see the analysis of Desjardins *et al.* (2001) for an insulating mantle). The friction term is important in the atmosphere, where the wind speed can be measured at different heights above the lower surface, but its strength is difficult to estimate at the fluid core boundaries. In addition, its role may be taken over in the core case by the electromagnetic force. We shall see indeed that electromagnetic coupling with the mantle appears as a friction term in the equation of torsional waves (within our restrictive hypotheses on the distribution of mantle conductivity). Assuming that on short time scales inertia predominates over viscous friction, the equation for the rapidly varying part $\mathbf{v_g}$ of the geostrophic velocity $\mathbf{u_g}$ is

$$\rho H \oint \frac{\partial \mathbf{v_g}}{\partial t} \cdot d\mathbf{\Gamma} = \int_{z_B}^{z_T} \left[\oint \left(\mathbf{j} \times \mathbf{B} - \alpha \rho \Theta \mathbf{g} - \rho \frac{d\mathbf{\Omega}}{dt} \times \mathbf{r} \right) \cdot d\mathbf{\Gamma} \right] dz \tag{3.12}$$

(anticipating possible fluctuations of the spin rate of the mantle and noting the position vector $\mathbf{r}$). In turn, a magnetic field $\tilde{\mathbf{b}}$, with the same characteristic time as the geostrophic velocity, is induced,

$$\frac{\partial \tilde{\mathbf{b}}}{\partial t} = \nabla \times (\mathbf{v_g} \times \mathbf{B}), \tag{3.13}$$

in the interior of the core, where the time changes of $\tilde{\mathbf{b}}$ are fast enough to make diffusion negligible. The Lorentz force seeks to return each geostrophic cylinder to its stable state, defined by Eqs (3.5) and (3.1), and torsional Alfvén waves arise.

I find it useful to derive once again (see Braginsky, 1970; Roberts and Soward, 1972) the torsional waves equation in order to discuss recent studies and to plan future works. Furthermore, Fearn and Proctor (1992) remarked that manipulations of the Lorentz force integral over a geostrophic cylinder that are very useful in the axisymmetrical case are not easily generalized to nonaxisymmetrical fields. Taking into account the nonaxisymmetrical component of the magnetic field at the core surface adds indeed a minor complication to the equation, which I think is best to be made explicit.

Braginsky (1970) derived the equations for the torsional Alfvén waves in the spherical case (Eq. (3.6) with $b = 0$). The buoyancy contribution to the left-hand side of Eq. (3.12) vanishes,

$$4\pi \rho s^2 z_T \frac{\partial(\omega_g + \Omega)}{\partial t} = \int_{-z_T}^{z_T} \oint (\mathbf{j} \times \mathbf{B})_\phi s \, d\phi \, dz, \tag{3.14}$$

where $\mathbf{v}_g = s\omega_g \mathbf{e}_\phi$ (see Eq. (3.6)). Braginsky assumed that a quasi-static state $(\mathbf{u}, \mathbf{B})$ exists and he considered $\tilde{\mathbf{b}}$ (see Eq. (3.13)) as a small perturbation. Indeed, condition Eq. (3.8), which is fulfilled in the quasi-static basic state, makes possible to linearize Eq. (3.14). Equation (3.13) does not hold at the boundary. There, a magnetic diffusion layer is set up to match the magnetic field induced in the core interior to the magnetic field in the mantle. It is convenient to study separately the contributions of the interior field $\tilde{\mathbf{b}}$ and of the diffusion layer field $\tilde{\mathbf{b}}_\lambda$ to Eq. (3.14). Equation (3.13) gives, in the interior of a spherical core,

$$\frac{\partial \tilde{\mathbf{b}}}{\partial t} = B_s s \frac{\partial \omega_g}{\partial s} \mathbf{e}_\phi - \omega_g \frac{\partial_1 \mathbf{B}}{\partial \phi}, \tag{3.15}$$

where ∂_1/∂_ϕ denotes differentiation with respect to ϕ holding $\mathbf{e}_s$, and $\mathbf{e}_\phi$ fixed. In turn, Eq. (3.15) gives the radial magnetic field at the bottom of the mantle $\tilde{b}_{mr}$ since the radial field is continuous across the magnetic diffusion layer. In a first stage, I suppose that the mantle is electrically insulating. Then, knowing $\tilde{b}_{mr}$ everywhere on the core surface, we deduce the two other components of the magnetic field $\tilde{\mathbf{b}}_m$ at bottom of the mantle. Finally, we have

$$\tilde{\mathbf{b}} + \tilde{\mathbf{b}}_\lambda = \tilde{\mathbf{b}}_m \quad \text{at } r = a. \tag{3.16}$$

This condition determines $\tilde{\mathbf{b}}_\lambda$.

Let us first study the contribution of the interior magnetic field to the right-hand side of Eq. (3.14). It is useful to remark that

$$\oint (\mathbf{j} \times \mathbf{B})_\phi \, d\phi = \frac{1}{s\mu_0} \oint \nabla \cdot (s \mathbf{B}_M B_\phi) \, d\phi, \tag{3.17}$$

where μ_0 is magnetic permeability, and $\mathbf{B}_M$ is the meridional magnetic field

$$\mathbf{B}_M = \mathbf{B} - B_\phi \mathbf{e}_\phi. \tag{3.18}$$

Equation (3.17) gives

$$\int_{-z_T}^{z_T} \oint (\mathbf{j} \times \mathbf{B})_\phi \, d\phi \, dz = \frac{1}{s^2 \mu_0} \frac{\partial}{\partial s} \left(s^2 \int_{-z_T}^{z_T} \oint B_s B_\phi \, d\phi \, dz \right)$$

$$+ \frac{a}{\mu_0 z_T} \left(\oint B_r B_\phi \, d\phi (s, z_T) + \oint B_r B_\phi \, d\phi (s, -z_T) \right). \tag{3.19}$$

Using $|\tilde{\mathbf{b}}| \ll |\mathbf{B}|$, we separate a volume term

$$I = \frac{1}{s^2 \mu_0} \frac{\partial}{\partial s} \left[s^2 \int_{-z_\mathrm{T}}^{z_\mathrm{T}} \oint (B_s \tilde{b}_\phi + B_\phi \tilde{b}_s) \, \mathrm{d}\phi \, \mathrm{d}z \right] \tag{3.20}$$

and a surface term

$$J = \frac{a}{\mu_0 z_\mathrm{T}} \oint \left[(B_r \tilde{b}_\phi + B_\phi \tilde{b}_r)(s, z_\mathrm{T}) + (B_r \tilde{b}_\phi + B_\phi \tilde{b}_r)(s, -z_\mathrm{T}) \right] \mathrm{d}\phi. \tag{3.21}$$

After taking the time derivation of Eq. (3.20), we can eliminate $\tilde{\mathbf{b}}$ through the use of Eq. (3.15),

$$\frac{\partial I}{\partial t} = \frac{4\pi}{s^2 \mu_0} \frac{\partial}{\partial s} \left(z_\mathrm{T} s^3 \frac{\partial \omega_\mathrm{g}}{\partial s} \{B_s^2\} \right), \tag{3.22}$$

where $\{B_s^2\}$ is a measure of the square of the s-component of the magnetic field averaged on each geostrophic cylinder:

$$\{B_s^2\}(s) = \frac{1}{4\pi z_\mathrm{T}} \int_{-z_\mathrm{T}}^{z_\mathrm{T}} \oint B_s^2 \, \mathrm{d}\phi \, \mathrm{d}z. \tag{3.23}$$

Anticipating the final result, I note J_λ the contribution of the magnetic field of the diffusion layer to the right-hand side of Eq. (3.14):

$$J_\lambda = \frac{1}{\mu_0} \int_{-z_\mathrm{T}}^{z_\mathrm{T}} \oint B_r \frac{\partial \tilde{b}_{\lambda\phi}}{\partial r} \, \mathrm{d}\phi \, \mathrm{d}z = \frac{1}{\mu_0} \int_{-z_\mathrm{T}}^{z_\mathrm{T}} \oint \frac{B_r}{\cos\theta} \frac{\partial \tilde{b}_{\lambda\phi}}{\partial z} \, \mathrm{d}\phi \, \mathrm{d}z \tag{3.24}$$

or

$$J_\lambda = \frac{a}{\mu_0 z_\mathrm{T}} \oint \left[(B_r \tilde{b}_{\lambda\phi})(s, z_\mathrm{T}) + (B_r \tilde{b}_{\lambda\phi})(s, -z_\mathrm{T}) \right] \mathrm{d}\phi. \tag{3.25}$$

By Eq. (3.16), we finally obtain

$$J + J_\lambda = \frac{a}{\mu_0 z_\mathrm{T}} \oint \left[(B_r \tilde{b}_{\mathrm{m}\phi} + B_\phi \tilde{b}_{\mathrm{m}r})(s, z_\mathrm{T}) + (B_r \tilde{b}_{\mathrm{m}\phi} + B_\phi \tilde{b}_{\mathrm{m}r})(s, -z_\mathrm{T}) \right] \mathrm{d}\phi. \tag{3.26}$$

With this expression, the equation for torsional Alfvén waves in a full sphere enclosed in an electrically insulating mantle is now complete. I write it below, Eq. (3.28), allowing for a thin layer of conducting material at the bottom of the mantle.

Studies of the Alfvén torsional waves riding inside the core may eventually lead to an assessment of the strength of the different torques acting at the CMB (Buffett, 1998). In particular, the electromagnetic torque has been thoroughly investigated. It is an obvious candidate as the damping mechanism of the waves whilst the viscous torque is usually neglected on the basis of its long time scale. Zonal motions at the core surface shearing an axisymmetrical magnetic field $\mathbf{B}_\mathrm{M}(r, \theta)$ induce meridional electrical currents $\mathbf{j}_\mathrm{M}(r, \theta)$ at the bottom of the conducting mantle. The resulting Lorentz force $\mathbf{j}_\mathrm{M} \times \mathbf{B}_\mathrm{M}$ is directed along $\mathbf{e}_\phi$ and exerts an axial torque on the mantle (see also Section 5). Thus, it is possible that a model of the Alfvén torsional waves, even one including only the interaction with the axisymmetrical part of the quasi-static magnetic field, yields an efficient electromagnetic torque acting on the mantle. I suppose that there is a thin conducting layer of conductivity $\sigma_\mathrm{m} \exp(-r/\Delta)$ at the bottom of the mantle. Following the considerations alluded to in the Introduction, I take the conductance of the layer $\sigma_\mathrm{m} \Delta(\theta, \phi)$ as laterally varying. The azimuthal magnetic field at the core surface is now $(\tilde{b}_{m\phi} + \tilde{b}_{\Delta\phi})$, where $\tilde{b}_{\Delta\phi}$ denotes the azimuthal field induced by the shear at the core surface, the notation $\tilde{\mathbf{b}}_m$ being saved for the magnetic field at bottom of the insulating volume inside the mantle. Assuming that $(\sigma_m\Delta \ll \sigma_c\delta_\lambda)$, where δ_λ is the thickness of the diffusion layer, and that $\Delta \ll \delta_H$, where δ_H is the length scale of variation of B_r at the core surface, we obtain by continuity of the electrical field parallel to the boundary

$$-\frac{\tilde{b}_{\Delta\phi}|_{r=a}}{\mu_0\sigma_m\Delta} = s\omega_g B_r. \tag{3.27}$$

Finally, the equation for Alfvén torsional waves is

$$\begin{aligned}
\rho s z_T \frac{\partial^2(\omega_g + \Omega)}{\partial t^2} &= \frac{1}{s^2\mu_0}\frac{\partial}{\partial s}\left(z_T s^3 \frac{\partial\omega_g}{\partial s}\{B_s^2\}\right) \\
&\quad - \frac{as}{4\pi z_T}\left(\oint \sigma_m\Delta B_r^2\,d\phi\,(s, z_T) + \oint \sigma_m\Delta B_r^2\,d\phi\,(s, -z_T)\right)\frac{\partial\omega_g}{\partial t} \\
&\quad + \frac{a}{4\pi\mu_0 z_T}\oint\left[\left(B_r\frac{\partial\tilde{b}_{m\phi}}{\partial t} + B_\phi\frac{\partial\tilde{b}_{mr}}{\partial t}\right)(s, z_T)\right. \\
&\quad \left. + \left(B_r\frac{\partial\tilde{b}_{m\phi}}{\partial t} + B_\phi\frac{\partial\tilde{b}_{mr}}{\partial t}\right)(s, -z_T)\right]d\phi. \tag{3.28}
\end{aligned}$$

In the axisymmetrical case, there is no azimuthal magnetic field at the core surface and the last term disappears. Braginsky (1970) suggested that this term can also be neglected on the ground that, at the core surface, in his geodynamo model (Braginsky, 1964), the nonaxisymmetrical part of the magnetic field is small compared to the axisymmetrical part. In the general case, whilst an expression of $\partial\tilde{b}_{mr}|_{r=a}/\partial t$ as a function of ω_g is directly obtained from Eq. (3.15), the determination of $\partial\tilde{b}_{m\phi}|_{r=a}/\partial t$ necessitates an integration over the entire core surface. Finally, operating with $4\pi\int_0^a s^2\,ds$ on Eq. (3.28) yields the time derivative of the torque budget, and as a consequence the equation that determines $d\Omega/dt$; the last term of Eq. (3.28) does not contribute. Of course, the torsional Alfvén waves have larger amplitude where $\{B_s^2\}(s)$ is weak. The Taylor's condition (3.5) is obeyed on time scales long compared to the period of the torsional waves, which is of the order of

$$\tau_{TA} = \frac{(\mu_0\rho)^{1/2}a}{\{B_s^2\}^{1/2}}, \tag{3.29}$$

where the denominator loosely refers to a typical value of $\{B_s^2\}(s)$. Braginsky adjusted this parameter to recover the characteristic time scale of the LOD variations, which is 60 years in his opinion. The frequency $\varpi = 2\pi\tau_{TA}^{-1}$ of the waves does not enter Eq. (3.28). Its introduction represents a important simplification in the writing of the equations only when the model includes a solid and conducting inner core. Then, we need to know the penetration depth in the inner core to avoid solving the induction equation there. Assuming that the time scale of the slow and large scale motions governed by Eq. (3.1) is of the order of the period $\tau_{MC} = \tau_d/\Lambda$ of the MC-waves (see the Introduction), we can check the consistency of the approach:

$$\frac{\tau_{TA}}{\tau_{MC}} = \left(\frac{B}{B_s}\right)\frac{1}{\tau_A\Omega}, \tag{3.30}$$

where τ_A is the period of the Alfvén waves that would exist inside the core in the absence of rotation (given by the expression (3.29) with B substituted for B_s). Here, in a diffusionless situation, magnetic and rotation effects are compared by the very small parameter $(\tau_A \Omega)^{-1}$. This result validates the assumption that the waves that arise when condition (3.5) is not satisfied consist of geostrophic motions.

Modelling of the torsional Alfvén waves has also been encouraged by the successful interpretation of core angular momentum changes as the result of acceleration of the geostrophic motions (Jault *et al.*, 1988; Jackson *et al.*, 1993; Hide *et al.*, 2000; Pais and Hulot, 2000). Models of zonal core surface velocities symmetrical with respect to the equatorial plane $u_\phi(\theta, t)$ ($\theta \leq \pi/2$ is colatitude) have been extracted from models of time-dependent core surface motions obtained after the inversion of secular variation data (in the mantle reference frame). In turn, $u_\phi(\theta, t)$ has been assimilated to $v_g(s)$. An estimate of time changes of core angular momentum $\mathcal{A}_c$ follows

$$\frac{d\mathcal{A}_c}{dt} = 4\pi\rho \frac{d}{dt} \int_0^a s^3 z_T[\omega_g(s) + \Omega]ds \tag{3.31}$$

and has been used to check, with LOD data, that core and mantle form a closed system on the decadal timescale:

$$\frac{d}{dt}(\mathcal{A}_c + \mathcal{A}_m) = 0, \tag{3.32}$$

denoting the mantle angular momentum by $\mathcal{A}_m$. However, the rapidly varying core surface motions that have been left out throughout the modelling remain mysterious.

There have been a few attempts to solve Eq. (3.28) numerically. In his pioneering study, Braginsky (1970) inferred the geometry of the meridional and quasi-static magnetic field from a plausible distribution of azimuthal electrical currents inside the core, making the condition (3.8) self-evident. His model included an insulating mantle and a conducting solid inner core, radius b, and angular velocity ω_i. He considered that the electromagnetic coupling between the torsional waves and the inner core is so efficient that

$$\forall s \leq b, \quad v_g(s) = s\omega_i \tag{3.33}$$

and solved Eq. (3.28) for $(v_g(s), s \geq b)$. Because, in his model, $\{B_s^2\}$ vanishes at $s = a$, the torsional oscillations are amplified in the equatorial region of the core. A few years ago, Buffett (1998) modified the Braginsky's model to include a possible gravitational torque between the inner core and the mantle and relinquished Eq. (3.33). I defer the discussion of this gravitational torque to the Section 4 but Buffett considered also an electromagnetic torque acting on a conducting mantle, which interests us here. He relied on Eq. (3.28) simplified as it befits the axisymmetrical case. He noticed (his Eq. (3.32)) that

$$\frac{\partial \omega_g}{\partial s} = -\mu_0\sigma_m\Delta\frac{\partial \omega_g}{\partial t} \quad \text{at } s = a, \tag{3.34}$$

when $B_r^2 \neq 0$ at the equator. Buffett used a numerical solution of the geodynamo equations (Kuang and Bloxham, 1999) as a substitute for a static basic state solution of Eq. (3.1) This does not, perhaps, represents an improvement on the initial state used by Braginsky since, very likely, condition (3.8) is violated. As in the Braginsky's study and for the same reason, the oscillations are confined to the equatorial region; the boundary condition (3.34) does not constrain the solutions (figure 1 of Braginsky (1970) and figure 6 of Buffett (1998)), even for an insulating mantle, because $B_r^2|_{\theta=\pi/2} = 0$, in both models. The amplitude of the waves is chosen to match the characteristic amplitude of core surface motions inferred from SV data. Buffett found that a mantle conductance of 10^8 S and a radial magnetic field at the CMB of uniform rms strength 5×10^{-4} T would make the electromagnetic torque strong enough to couple the Alfvén torsional waves with the mantle and explain LOD data. As Braginsky had forecast, this torque would be associated with heavy damping of the torsional waves. This modelling of the electromagnetic torque is however not conclusive because its success hinges on both a strong radial magnetic field B_r and a vanishing magnetic field B_s, in the equatorial region. The latter feature is required to enable amplification of the torsional waves, which augments the coupling. The two fields should merge at $s = a$ though.

Zatman and Bloxham (1997b, 1998) pioneered recently an inverse modelling of the torsional waves. As in core angular momentum studies, a model of $u_\phi(\theta, t)$, for 1900–1990, has been transformed into a model of $v_g(s, t)$. The latter has then been converted into one or two torsional waves of definite imaginary frequency ϖ. In turn, the waves are inverted for models of $\{B_s^2\}$ and of an ad hoc 'friction' coefficient at the CMB. The model incorporates nonaxisymmetrical effects. However, Zatman and Bloxham did not use Eq. (3.28). Instead, they replaced $\tilde{\mathbf{b}}_m|_{r=a}$, the magnetic field at the bottom of the mantle, by $\tilde{\mathbf{b}}|_{r=a}$, the magnetic field induced in the core interior, in Eq. (3.28), and then they determined $\tilde{\mathbf{b}}$ by Eq. (3.15). That amounts to omit the magnetic diffusion layer. Zatman and Bloxham tried to get round this difficulty by introducing a coefficient α, $0 \le \alpha \le 1$ multiplying the surface term to mimic the influence of an insulating mantle, as found by Braginsky in the axisymmetrical case. But, as shown above, the surface term $J + J_\lambda$ does not vanish, even with an insulating mantle, in the nonaxisymmetrical case. Furthermore, in their model, α multiplies only a part of the surface term. The introduction of the coefficient α amounts indeed to replace $B_r \tilde{b}_\phi$ in the surface term by $(\alpha s B_s + z B_z) \tilde{b}_\phi / a$. Thus, taking $\alpha = 0$ does not suffice to cancel the surface term. It is not clear how the results depend on the incorrect substitution of $\tilde{\mathbf{b}}|_{r=a}$ for $\tilde{\mathbf{b}}_m|_{r=a}$. The most striking result is the abrupt increase of $\{B_s^2\}$ with colatitude θ at about $\theta = 60°$.

Equation (3.28) gives also the response of the core to changes in the rotation rate of its container. Taking, as an example, the estimates of $\sigma_m \Delta$ and $B_r^2|_{r=a}$ obtained from nutation studies, we calculate a spin-up time of the order of 50 years, reduced to 5 years for the outer geostrophic cylinders representing one-tenth of core angular momentum. These values, which are probably on the lower side, are compatible with the observations reviewed in Section 2.

I consider now nonaxisymmetrical topography at the CMB. Geostrophic contours are neither circular nor planar: any topography symmetrical with respect to the equatorial plane distorts the contours in the s-direction whilst antisymmetrical topography bends them in the z-direction. Yet, there are very few studies bearing on the coupled equations (3.12) and (3.13) when the contours are not circular. Anufriyev and Braginsky (1977) assumed that a zonal velocity $v_\phi(s, z, t)$, taken as representative of torsional waves, is present in an axisymmetrical reference state and investigated topographical effects as a perturbation only. They supposed also that the magnetic field $\mathbf{B_M}$, that is responsible for the torsional waves in the first place, can be neglected, as far as the perturbations caused by topography are concerned, in comparison with a zonal magnetic field $B_\phi(s, z)\mathbf{e}_\phi$. Their study aimed at evaluating the pressure torque that acts on the casing:

$$\int_\Sigma p(\mathbf{r} \times \mathbf{n}) \, dS. \qquad (3.35)$$

In the event, topographical effects were found to be negligible. Braginsky (1998) followed up this study with another, in a plane layer approximation, including the consequence of a possible density stratification of the upper layers of the core. He found then that the pressure forces, at the CMB, exert a significant torque on the mantle. Topographical effects are indeed amplified because of the impenetrable boundary between the top layer and the core interior. The difficulty with the Anufriyev and Braginsky approach is its artificial character. In the actual problem, $\mathbf{v}_g$ flows along geostrophic contours, not along circular contours, and it is not perturbed by the aspherical boundary. Their study applies only if there is a core process, necessarily different from the torsional oscillations, that produce a zonal velocity $v_\phi(s, z, t)$ (with a decadal time scale) that does not follow the geostrophic contours. The work can be summarized as a study of forced Rossby waves in the presence of a magnetic field.

Unfortunately, the important point, that is, the substitution of Eq. (3.12) to (3.14) when the contours are not circular has inspired very few studies, in the context of core–mantle coupling, at least. At first order ε in the topography:

$$r|_{\text{CMB}} = a[1 + \varepsilon h(\theta, \phi)], \quad h = O(1). \tag{3.36}$$

There is now a contribution from the nonaxisymmetrical part of the Lorentz force and from the buoyancy term. In the axisymmetrical case, multiplying Eq. (3.14) by s and integrating from $s = 0$ to $s = a$ readily gives an expression of the torque acting on the core, which involves only the Lorentz force accelerating the rotation of the geostrophic cylinders. The situation is more intricate in the nonaxisymmetrical case. Equation (3.12) does not yield a torque budget (Fearn and Proctor, 1992) and it does not indicate what forces exert a torque on the mantle. A somewhat artificial model incorporating an insulating mantle and a spherically symmetric gravity field is then an useful guide. Both gravity and electromagnetic torque vanish and yet the total core angular momentum, carried by geostrophic motions, can change. It turns out that, in this case, the pressure exerted on the CMB causes the angular momentum exchange between core and mantle even though the pressure gradient does not enter Eq. (3.12). Omitting entirely electromagnetic forces within the core and assuming $\varepsilon \ll 1$, Jault *et al.* (1996) found that the expressions giving respectively the action of the gravity force on each geostrophic cylinder $\mathcal{C}$ (see Eq. (3.12)) and the pressure torque exerted on the mantle at the rim of $\mathcal{C}$ are equivalent. In the general case (with magnetic forces), we have to rely on Eq. (3.12) to infer the time changes of the geostrophic velocity.

In conclusion, the theory of Alfvén torsional waves has given us, by far, the most robust link between theories of Earth's dynamo and observations. It explains most of the decade variations in the LOD and, at least, some of the rapid variations of the Earth's magnetic field. Even if the first efforts of data assimilation have been promising, there is scope for further studies. An interesting step would be to construct a model of a quasi-static magnetic field satisfying either Eq. (3.8) or (3.5) and compatible with a model of the magnetic field at the Earth's surface. Through direct modelling, the hypotheses of topographic and electromagnetic coupling could then be tested. There is still no dynamical study of topographic coupling relevant to the Earth's core problem. In this context, the topography is important inasmuch it determines the geostrophic contours. It enters the model only through Eq. (3.12), whilst magnetic field induction can be calculated in a spherical geometry. Concerning electromagnetic coupling, it is likely that it implies strongly damped torsional waves (Bloxham, 1998; Zatman and Bloxham, 1998). Relying on Eq. (3.28), this can be quantified. Finally, more sophisticated models of the geostrophic circulation inverted from magnetic field data and Eq. (3.32) are also possible. According to Zatman and Bloxham (1998), the decay times of the waves are at most of the order of their periods. As a result, the assumption that the geostrophic velocity, inside the core, can be represented as the superposition of a few waves with definite frequency appears questionable. This hypothesis is not necessary either, since the frequency of the waves does not enter Eq. (3.28).

4. Deviations from axisymmetry of the solid inner core shape

If the two boundaries enclosing the fluid core, of radius respectively $r \simeq b$ and $r \simeq a$, are not perfectly axisymmetrical, there are regions void of geostrophic contours in the vicinity of $s = b$ and $s = a$. Near $s = b$, the crucial surface is the cylinder Π^b, which is parallel to the z-axis and touches the outer rim of the inner core, at a location denoted $z^b(\phi)$. The cylinder intersects the outer boundary at $z = z_T^b(\phi)$ and $z = z_B^b(\phi)$. Just inside Π^b and above the inner core, there is a geostrophic cylinder $C(H^{b,i})$ of constant height $H^{b,i} = \min(z_T^b(\phi) - z_B^b(\phi))$ but there are no geostrophic contours between $C(H^b)$ and Π^b. In the same way, there is a geostrophic cylinder $C(H^{b,o})$ of constant height $H^{b,o} = \min(z_T^b(\phi) - z_B^b(\phi))$ just outside Π^b but no geostrophic contours between Π^b and $C(H^{b,o})$. In the absence of closed contours of constant height, the role of the geostrophic motions is usually taken over by low-frequency z-independent inertial waves (Greenspan, 1968), known as 'Rossby waves'. Outside Π^b, the height variation of fluid columns circling around the inner core is of the order of h/a and the frequency of the Rossby waves is thus of order $(h/a)\Omega$. This is comparable with the frequency of the torsional Alfvén waves. On the other hand, the special regions, where no geostrophic contours exist, represent a small portion of the fluid volume. This explains that they do not play a role in the model of Buffett (1996), which I outline now.

Buffett studied the coupling of the rotation of the inner core to the torsional waves in the fluid outer core. He supposed that the inner core surface Σ_b is an equipotential surface of the Earth's gravity field. The hydrostatic pressure is indeed constant, in the fluid outer core, on these equipotential surfaces and the pressure determines the freezing point of iron. Because of density anomalies in the mantle, the Earth's gravity field is probably not axisymmetrical and neither is the inner core surface (see Introduction). Buffett studied the free mode of axial rotation of the solid inner core in this configuration. When the inner core is rotated from its equilibrium position, an archimedean force arises in response to the misalignment between the inner core and the mantle. The gravity torque acting on the fluid outer core from the mantle is compensated by the pressure torque acting from the inner core whilst the net torque acting on the inner core is nonzero because of the density jump at Σ_b. Finally, Buffett found that the period of this rotation eigenmode is of the order of a few years. This is small compared to the period of the torsional waves. Hence, according to this study, the inner core is locked to the mantle. In addition to all the other torques acting on the mantle, the fluid outer core and the mantle may thus be coupled through the electromagnetic torque acting between the fluid and solid cores. This mechanism can theoretically be tested from seismological observations of the inner core rotation (Buffett and Creager, 1999). Viscous deformation of the inner core in response to gravity and pressure forces may loosen the grip of the mantle on the inner core. Assuming a newtonian rheology for the inner core (neglecting elastic deformation), Buffett investigated different values of the viscous relaxation time of the inner core τ_v. He concluded, using the rather high value of inner core topography found by Forte and Peltier (1991), that gravitational coupling between the inner core and the mantle remains important within a wide range of values of τ_v. This mechanism involves electromagnetic coupling between the inner and outer cores and thus implies some damping of the torsional waves (Buffett, 1996). Finally, if the gravitational torque is important, the ensemble inner core–mantle is coupled to the fluid core through torsional waves of period τ_{TA}. This is once more compatible with the results of Section 2 about the characteristic timescale of the coupling mechanism between outer core and mantle. Gravitational coupling is an attractive mechanism to explain changes in LOD. Assessment of its importance awaits renewed confrontation with geomagnetic and seismic data (see the promising study of Vidale *et al.* (2000)) together with improved modelling of topographic and electromagnetic coupling at the CMB.

5. Kinematic modelling

Considering core–mantle coupling as a consequence of torsional Alfvén waves, we presume that we have attained a good understanding of core dynamics. However, torsional waves do not explain all the rapid changes of the Earth's magnetic field. Furthermore, I conclude from the above review that topographic and electromagnetic coupling have not yet been satisfactorily incorporated in models of torsional waves. These weaknesses of a fully dynamic approach justify the less ambitious kinematic studies that I report now. Neglecting diffusion, models of core surface motions $\mathbf{u}|_{r=a}$ have been inverted from the radial component of the induction equation

$$\frac{\partial B_r}{\partial t} = -\nabla_H \cdot (\mathbf{u} B_r). \tag{3.37}$$

They include large-scale, nongeostrophic flows that change on a decadal time scale. In the present state of core studies, these motions appear enigmatic. Leaving aside the question of their origin, a kinematic modelling of core–mantle coupling is nevertheless possible. It consists in investigating the different torques acting on the mantle that are associated with these flows. In this context, two mechanisms have been particularly studied. First, electrical potential differences are set up at the core surface. Thus, electrical currents flow in the mantle if it is not a perfect insulator and a Lorentz force exerts a torque on the mantle. Second, a pressure is associated to the motions and is applied also on the solid mantle.

Roberts (1972) long ago gave the general formulation of the electromagnetic torque. The approximate expression derived in Section 3 applies to the case of zonal motions at the core surface and thin electrically conducting layer at the bottom of the mantle but the conclusions are not radically altered in less specific cases. All authors have assumed that the magnetic field induced in the mantle is a perturbation of the magnetic field generated by core motions (Benton and Whaler, 1983). In other words, the electrical potential differences are not short-circuited by the conducting mantle. It is convenient to write again the continuity of the electrical field tangent to the boundary across this surface. The motions $\mathbf{u}|_{r=a}$ enter Eq. (3.37) only through the term

$$\mathbf{u} B_r = \nabla_H \Psi + \nabla \times (\mathbf{r} \Phi), \quad V = a\Phi, \tag{3.38}$$

where V is electrical potential and Ψ and Φ are two scalar fields on the core surface. The electrical potential can be calculated in the mantle from its value at the CMB and the equation for electrical charge conservation

$$\nabla \cdot (\sigma_m \nabla V) = 0. \tag{3.39}$$

From Eq. (3.37), the calculation of Ψ is straightforward. This term arises because of time changes of the magnetic field permeating the conducting mantle; it is associated with electrical currents that can be directly calculated from SV models also. However, the magnetic field, in the mantle, is largely axisymmetrical. Taking into account $\mathbf{B}_M(r, \theta)$ only, the azimuthal electrical currents induced by the time changes of $\mathbf{B}_M(r, \theta)$ do not enter the expression of the azimuthal Lorentz force $F_{B,\phi} = (\mathbf{j} \times \mathbf{B})_\phi$ that acts as a torque on the mantle. On the other hand, an electrical potential $V(a, \theta)$ is set up by any zonal motion at the core surface (see Eq. (3.38)). The resulting magnetic force in the mantle,

$$\mathbf{F}_B = \sigma_m \nabla V(r, \theta) \times \mathbf{B}_M(r, \theta), \tag{3.40}$$

is directed along $\mathbf{e}_\phi$ and is very efficient to act as a torque on the mantle. Finally, the main part of the torque is determined by the unknown scalar Φ, which is almost not constrained by SV data (Jault and Le Mouël, 1991) because the magnetic field in the mantle is predominantly zonal. Holme (1998a) relied on this property to show, through inverse modelling, that there are models of $\mathbf{u}|_{r=a}$ that are compatible with geomagnetic observations and yield an electromagnetic torque that, taken in isolation, would explain LOD data, if the mantle is sufficiently conducting. Using annual means of the magnetic field published by the observatories complemented by the Bloxham and Jackson (1992) magnetic field model for the period 1900–1980, he found $\sigma_m \Delta = 10^8$ S as a reasonable minimum value for the mantle conductance (Holme, 1998b) to make the electromagnetic torque significant. This is about five times less than the value that was deemed necessary from direct modelling. Wicht and Jault (1999) based their investigation of the electromagnetic torque on Φ instead of $\mathbf{u}|_{r=a}$, as in Holme (1998a). The idea was to monitor precisely the uncertainty. There is indeed one single information on Φ:

$$\nabla \times (\mathbf{r}\Phi) = -\nabla_H \Psi \quad \text{when} \quad B_r = 0. \tag{3.41}$$

Knowing that we are interested only by Φ, we tried to give more weight to Eq. (3.41). However, we mainly confirmed the result of Holme (1998b) for the minimum value of $\sigma_m \Delta$ and we were not able to increase it despite the refined constraint (3.41). Finally, the consequences of possible lateral variations in the electrical conductivity at the bottom of the mantle have been recently investigated (Holme, 2000; Wicht and Jault, 2000). It turns out that the contributions of the different regions at the core surface are simply weighted by the conductance of the mantle nearby.

If the CMB is aspherical (see Eq. (3.36) defining the small parameter ε), the moment of the pressure force acting on the mantle from the core may be nonzero (Hide, 1969; see Eq. (3.35)). Neglecting the Lorentz force at the core surface, a kinematic approach is possible (Hide, 1989; Jault and Le Mouël, 1989). The pressure is calculated in the spherical approximation

$$\mathbf{u} = \mathbf{u}_0 + \varepsilon \mathbf{u}_1; \quad p = p_0 + \varepsilon p_1. \tag{3.42}$$

At the core surface, and at zeroth order in ε, Eqs (3.1) and (3.3) are transformed into:

$$2\rho(\mathbf{\Omega} \times \mathbf{u}_0) = -\nabla p_0 - \alpha\theta\mathbf{g}, \quad \mathbf{u}_0 \cdot \mathbf{e}_r = 0. \tag{3.43}$$

According to the kinematic approach that I adopt here, the velocity $\mathbf{u}_0$ is known from Eq. (3.37). The pressure p_0 is then obtained from the horizontal components of Eq. (3.43). The pressure torque arises at first order in ε. Omitting other possible torques for the sake of simplicity (they can be reinstated later) and operating with $\mathbf{e}_z \cdot \int \mathbf{r} \times$ on the equation of motion, I obtain

$$\int \mathbf{e}_\phi \cdot s\rho \frac{\partial \mathbf{u}_0}{\partial t} dV = -\mathbf{e}_z \cdot \left(\int_{\text{CMB}} (\mathbf{r} \times p_0\mathbf{n}) dS + \int_{r=a} (\mathbf{r} \times p_1\mathbf{e}_r) dS \right). \tag{3.44}$$

The contribution of the first-order pressure vanishes and a model of p_0, together with a model of CMB topography, suffices to calculate the pressure torque. In order to make the derivation consistent, the inertial term has to enter the equation of motion at the order ε also. This means that the angular momentum carried by core motions changes on the characteristic time $(\varepsilon\Omega)^{-1}$. Decadal time scales are obtained for topographies of a few hundreds of meters. More detailed studies are frustrated by the lack of models of the topography at the CMB. In addition, the pressure force with short lengthscale along the core surface is not constrained by SV data but may nevertheless exert a significant torque on the mantle. These two shortcomings explain that there have been no attempts at calculating a time series of the pressure torque from SV data. If anything, the topographic torque inferred from kinematic modelling is too potent.

6. Concluding remarks

As we have seen, the main limitation of the models of torsional Alfvén waves is their inability to explain some rapid changes in the Earth's magnetic field that are currently attributed to nongeostrophic large scale motions, themselves of unknown origin. It is, of course, possible to devise explanations for such global flows. Braginsky (1993) noticed that if the upper core were chemically stratified, it would support fast waves of long lengthscale. Another plausible way to reconcile the theory of torsional waves with observations is to interpret the rapid variations of the Earth's magnetic field as the result of short lengthscale motions. As a matter of fact, an important part of the SV models can already be well explained by steady or slowly varying large-scale core surface motions, which are expected from dynamo modelling. Rapidly varying flows are often penalized in models of velocity at the core surface, but short- and long-lengthscale motions are treated in the same way whilst rapid changes of the short lengthscale component are much more acceptable from a physical standpoint. Note also that the small-scale components, of harmonic degree $l \geq 10$, have been underestimated in the magnetic field model that has been the reference for the last 15 years (Bloxham and Jackson, 1992). This has made the contribution of small-scale motions to the secular variation potentially less important. Finally, a temporal norm has been minimized in the inversion of the magnetic field model itself. It does not distinguish either between short- and long-lengthscale components. All these assumptions may well have conspired to mistake rapid variations of the small-scale flow for large-scale motions. Before the satellite era that we are now entering, it was difficult to test more sophisticated a priori models of the magnetic field and of the surface core flows for want of data. Now, observations of geomagnetic field, with smaller lengthscales and shorter time scales become available. Thus, the modelling of core surface motions will require less regularization.

Finally, there is an interesting theoretical question still pending. It transpires that the amplitude of the torsional wave velocity matches up to the amplitude of the quasi-static velocity field, which is part of the dynamo process. This coincidence calls for an explanation, which may involve a discussion of the coupling with the mantle. In the view expounded here, excitation of the Alfvén torsional waves stems from the slow evolution of the quasi-static state. On the other hand, the coupling mechanism between the core and the mantle is not immaterial to the amplitude of the torsional waves since electromagnetic coupling with either the inner core gravitationally locked to the mantle or directly with the mantle probably entails strong damping. The situation is different in the case of topographic coupling.

Acknowledgements

Jacques Hinderer, Véronique Dehant and Olivier de Viron have kindly informed me about very recent, sometimes unpublished, studies. Sonny Mathews has sent me a preprint of his latest paper before its publication. Alexandra Pais has trusted me with her notes about the torsional oscillations in the core and an electronic mail from Stephen Zatman has also enlightened me. Fritz Busse and Daniel Brito have made important suggestions to improve the manuscript.

References

Abarca del Rio, R., "The influence of global warming in Earth rotation speed," *Ann. Geophysicae* **17**, 806–811 (1999).

Abarca del Rio, R., Gambis, D. and Salstein, D. A., "Interannual signals in length of day and atmospheric angular momentum," *Ann. Geophysicae* **18**, 347–364 (2000).

Anufriyev, A. P. and Braginsky, S. I., "Effects of irregularities of the boundary of the Earth's core on the speed of the fluid and on the magnetic field, iii," *Geomag. Aeron.* **17**, 492–496 (1977).

Bassom, A. P. and Soward, A. M., "Localised rotating convection induced by topography," *Physica D* **97**, 29–44 (1996).

Bell, P. I. and Soward, A. M., "The influence of surface topography on rotating convection," *J. Fluid Mech.* **313**, 147–180 (1996).

Benton, E. R. and Whaler, K. A., "Rapid diffusion of the poloidal geomagnetic field through the weakly conducting mantle: a perturbation solution," *Geophys. J. R. Astr. Soc.* **75**, 77–100 (1983).

Bloxham, J., "Dynamics of angular momentum in the Earth's core," *Annu. Rev. Earth Planet. Sci.* **26**, 501–517 (1998).

Bloxham, J. and Jackson, A., "Time-dependent mapping of the magnetic field at the core–mantle boundary," *J. Geophys. Res.* **97** (B13), 19537–19563 (1992).

Braginsky, S. I., "Self-excitation of a magnetic field during the motion of a highly conducting fluid," *Sov. Phys. JETP* **20**, 726–735 (1964).

Braginsky, S. I., "Torsional magnetohydrodynamic vibrations in the Earth's core and variations in day length," *Geomag. Aeron.* **10**, 1–8 (1970).

Braginsky, S. I., "MAC oscillations of the hidden ocean of the core," *J. Geomag. Geoelectr.* **45**, 1517–1538 (1993).

Braginsky, S. I., "Magnetic Rossby waves in the stratified ocean of the core, and topographic core–mantle coupling," *Earth Planets Space* **50**, 641–649 (1998).

Buffett, B. A., "Constraints on magnetic energy and mantle conductivity from the forced nutations of the Earth," *J. Geophys. Res.* **97**, 19581–19597 (1992).

Buffett, B. A., "A mechanism for decade fluctuations in the length of day," *Geophys. Res. Lett.* **23**, 3803–3806 (1996).

Buffett, B. A., "Free oscillations in the length of day: inferences on physical properties near the core–mantle boundary," In: *The Core–Mantle Boundary Region* (Eds M. Gurnis, M. E. Wysession, E. Knittle and B. A. Buffett) Geodynamics Series, **28**, pp. 153–165, AGU (1998).

Buffett, B. A. and Creager, K. C., "A comparison of geodetic and seismic estimates of inner core rotation," *Geophys. Res. Lett.* **26**, 1509–1512 (1999).

Buffett, B. A., Garnero, E. J. and Jeanloz, R., "Sediments at the top of Earth's core," *Science* **290**, 1338–1342 (2000).

Celaya, M. A., Wahr, J. M. and Bryan, F. O., "Climate-driven polar motion," *J. Geophys. Res.* **104**, 12813–12829 (1999).

Defraigne, J. A., Dehant, V. and Wahr, J., "Internal loading of an inhomogeneous compressible Earth with phase boundaries," *Geophys. J. Int.* **125**, 173–192 (1996).

Desjardins, B., Dormy, E. and Grenier, E., "Instability of Ekman–Hartmann boundary layers, with application to the fluid flow near the core–mantle boundary," *Phys. Earth Planet. Inter.* **123**, 15–26 (2001).

Dickey, J. O. and Hide, R., "Earth's variable rotation," *Science* **253**, 629–637 (1991).

Dickey, J. O., Marcus, S. L. and de Viron, O., "The Earth angular momentum budget on subseasonal timescales: exchange among the solid Earth, atmosphere and ocean subsystems," *EOS Trans. AGU* **81** (48), G61B–25, Fall meet. suppl. (2000).

Dickman, S. R., "Estimates of core–mantle torques from rotational and gravitational data," *Geophys. J. Int.* **144**, 532–538 (2001).

Dickman, S. R. and Nam, Y. S., "Constraints on Q at long periods from Earth's rotation," *Geophys. Res. Lett.* **25**, 211–214 (1998).

Dobson, D. P. and Brodholdt, J. P., "The electrical conductivity and thermal profile of the earth's mid-mantle," *Geophys. Res. Lett.* **27**, 2325–2328 (2000).

Eubanks, T. M., "Variations in the orientation of the Earth," In: *Contributions of Space Geodesy to Geodynamics: Earth Dynamics* (Eds D. E. Smith and D. L. Turcotte) Geodynamics Series, **24**, pp. 1–54, AGU (1993).

Eubanks, T. M., Steppe, J. A., Dickey, J. O. and Callahan, P. S., "A spectral analysis of the Earth–atmosphere angular momentum budget," *J. Geophys. Res.* **90**, 5385–5404 (1985).

Fearn, D. R., "Hydromagnetic flow in planetary cores," *Rep. Prog. Phys.* **61**, 175–235 (1998).

Fearn, D. R. and Proctor, M. R. E., "Magnetostrophic balance in non-axisymmetric, non-standard dynamo models," *Geophys. Astrophys. Fluid Dynam.* **67**, 117–128 (1992).

Fearn, D. R., Roberts, P. H. and Soward, A. M., "Convection, stability and the dynamo," In: *Energy Stability and Convection* (Eds G. P. Galdi and B. Straughan) Pitman Research Notes in Mathematics Series, **168**, pp. 60–324, Longman Scientific & Technical, New York (1988).

Forte, A. and Peltier, R., "Viscous flow model of global geophysical observables 1. forward problems," *J. Geophys. Res.* **96**, 20131–20159 (1991).

Garnero, E. J., "Heterogeneity of the lowermost mantle," *Annu. Rev. Earth Planet. Sci.* **28**, 509–537 (2000).

Gibert, D., Holschneider, M. and Le Mouël, J.-L., "Wavelet analysis of the Chandler wobble," *J. Geophys. Res.* **103**, 27069–27089 (1998).

Greenspan, H. P., *The Theory of Rotating Fluids*. Cambridge University Press, Cambridge (1968).

Gross, R. S., "The excitation of the Chandler wobble," *Geophys. Res. Lett.* **27**, 2329–2332 (2000).

Gurnis, M., Wysession, M. E., Knittle, E. and Buffett, B. A., (eds) *The Core–Mantle Boundary region*, Geodynamic series. **28** AGU (1998).

Gwinn, C. R., Herring, T. A. and Shapiro, I. I., "Geodesy by interferometry: studies of the forced nutations of the Earth. 2, interpretation," *J. Geophys. Res.* **91**, 4755–4765 (1986).

Hide, R., "Interaction between the Earth's liquid core and solid mantle," *Nature* **222**, 1055–1056 (1969).

Hide, R., "Fluctuations in the earth's rotation and the topography at the core–mantle interface," *Phil. Trans. R. Soc. Lond.* A **328**, 351–363 (1989).

Hide, R., Boggs, D. H., Dickey, J. O., Dong, D. and Gross, R. S., "Topographic core–mantle coupling and polar motion on decadal time-scales," *Geophys. J. Int.* **125**, 599–607 (1996).

Hide, R., Boggs, D. H. and Dickey, J. O., "Angular momentum fluctuations within the Earth's liquid core and torsional oscillations of the core–mantle system," *Geophys. J. Int.* **143**, 777–786 (2000).

Holme, R., "Electromagnetic core–mantle coupling I: explaining decadal changes in the length of day," *Geophys. J. Int.* **132**, 167–180 (1998a).

Holme, R., "Electromagnetic core–mantle coupling II: probing deep mantle conductance," In: *The Core–Mantle Boundary Region* (Eds M. Gurnis, M. E. Wysession, E. Knittle and B. A. Buffett) Geodynamics Series, **28**, pp. 139–151, AGU (1998b).

Holme, R., "Electromagnetic core–mantle coupling III: laterally varying mantle conductance," *Phys. Earth Planet. Inter.* **117**, 329–344 (2000).

Hulot, G., Le Huy, M. and Le Mouël, J.-L., "Influence of core flows on the decade variations of the polar motion," *Geophys. Astrophys. Fluid Dynam.* **82**, 35–67 (1996).

Jackson, A., "Time-dependency of tangentially geostrophic core surface motions," *Phys. Earth Planet. Inter.* **103**, 293–311 (1997).

Jackson, A., Bloxham, J. and Gubbins, D., "Time-dependent flow at the core surface and conservation of angular momentum in the coupled core–mantle system," In: *Dynamics of the Earth's Deep Interior and Earth Rotation* (Eds J. L. LeMouël, D. E. Smylie and T. Herring) Geophysical Monograph Series, **92**, pp. 97–107, AGU (1993).

Jault, D. and Le Mouël, J.-L., "The topographic torque associated with a tangentially geostrophic motion at the core surface and inferences on the flow inside the core," *Geophys. Astrophys. Fluid Dynam.* **48**, 273–296 (1989).

Jault, D. and Le Mouël, J.-L., "Exchange of angular momentum between the core and the mantle," *J. Geomag. Geoelectr.* **43**, 111–129 (1991).

Jault, D., Gire, C. and Le Mouël, J.-L., "Westward drift, core motions and exchanges of angular momentum between core and mantle," *Nature* **333**, 353–356 (1988).

Jault, D., Hulot, G. and Le Mouël, J.-L., "Mechanical core–mantle coupling and dynamo modelling," *Phys. Earth Planet. Inter.* **43**, 187–191 (1996).

Kuang, W. and Bloxham, J., "Numerical modeling of magnetohydrodynamic convection in a rapidly rotating spherical shell: weak and strong field dynamo action," *J. Comput. Phys.* **153**, 51–81 (1999).

Mc Carthy, D. and Luzum, D., "Possible existence of very low frequency periodic motion," *Geophys. J. Int.* **125**, 623–629 (1996).

Mathews, P. M., Herring, T. A. and Buffett, B. A., "Modeling of nutation-precession: new nutation series for nonrigid Earth, and insights into the Earth's interior," *J. Geophys. Res.* **107** (B4), 10.129/2001JB000390 (2002).

Neuberg, J., Hinderer, J. and Zürn, W., "Stacking gravity tide observations in Central Europe for the retrieval of the complex eigenfrequency of the nearly diurnal free wobble," *Geophys. J. R. Astr. Soc.* **91**, 853–868 (1987).

Neubert, T., Mandea, M., Hulot, G., von Frese, R., Friis-Christensen, E., Stauning, P., Olsen, N. and Risbo, T., "Ørsted satellite captures high-precision geomagnetic field data," *EOS Trans. AGU* **82** (7), 81, 87–88 (2001).

Pais, A. and Hulot, G., "Length of day decade variations, torsional oscillations and inner core superrotation: evidence from recovered core surface zonal flows," *Phys. Earth Planet. Inter.* **118**, 291–316 (2000).

Petersons, H. F. and Constable, S., "Global mapping of the electrically conductive lower mantle," *Geophys. Res. Lett.* **23**, 1461–1464 (1996).

Roberts, P. H., "Electromagnetic core–mantle coupling," *J. Geomag. Geoelectr.* **24**, 231–259 (1972).

Roberts, P. H. and Soward, A. M., "Magnetohydrodynamics of the Earth's core," *Annu. Rev. Fluid Mech.* **4**, 117–154 (1972).

Rosen, R. D., Salstein, D. A. and Wood, T. M., "Discrepancies in the Earth–atmosphere angular momentum budget," *J. Geophys. Res.* **95**, 265–279 (1990).

Schultz, A., Kurtz, R. D., Chave, A. D. and Jones, A. G., "Conductivity discontinuities in the upper mantle beneath a stable craton," *Geophys. Res. Lett.* **20**, 2941–2944 (1993).

Smith, M. L. and Dahlen, F. A., "The period and Q of the Chandler wobble," *Geophys. J. R. Astr. Soc.* **64**, 223–281 (1981).

Taylor, J. B., "The magneto-hydrodynamics of a rotating fluid and the earth's dynamo problem.," *Proc. R. Soc. Lond.* A **274**, 274–283 (1963).

Vidale, J. E., Dodge, D. A. and Earle, P. S., "Slow differential rotation of the earth's inner core indicated by temporal changes in scattering," *Nature* **405**, 445–448 (2000).

Wicht, J. and Jault, D., "Constraining electromagnetic core–mantle coupling," *Phys. Earth Planet. Inter.* **111**, 161–177 (1999).

Wicht, J. and Jault, D., "Electromagnetic core–mantle coupling for laterally varying mantle conductivity," *J. Geophys. Res.* **105**, 23569–23578 (2000).

Wu, X. and Wahr, J. M., "Effets of non-hydrostatic core–mantle boundary topography and core dynamics on Earth rotation," *Geophys. J. Int.* **128**, 18–42 (1997).

Xu, Y., McCammon, C. and Poe, B. T., "The effect of alumina on the electrical conductivity of silicate perovskite," *Nature* **282**, 922–924 (1998).

Zatman, S. and Bloxham, J., "The phase difference between length of day and atmospheric angular momentum at subannual frequencies and the possible role of core–mantle coupling," *Geophys. Res. Lett.* **24**, 1799–1802 (1997a).

Zatman, S. and Bloxham, J., "Torsional oscillations and the magnetic field within the Earth's core," *Nature* **388**, 760–763 (1997b).

Zatman, S. and Bloxham, J., "A one-dimensional map of B_s from torsional oscillations of the Earth's core," In: *The Core–Mantle Boundary Region* (Eds M. Gurnis, M. E. Wysession, E. Knittle and B. A. Buffett) Geodynamics Series, **28**, pp. 183–196, AGU (1998).

4 Geomagnetic reversals

Rates, timescales, preferred paths, statistical models, and simulations

Catherine G. Constable

Institute of Geophysics and Planetary Physics, Scripps Institution of Oceanography, University of California at San Diego, La Jolla, CA 92093-0225, USA

Paleomagnetic data on geomagnetic reversals are divided into two general categories: times of occurrence, and records of directional and/or intensity changes for transitions at individual locations. Despite considerable efforts expended in acquiring paleomagnetic reversal records, a detailed picture of the reversal process is still lacking, along with any means of clearly identifying when the magnetic field has entered a transitional state destined to lead to a reversal. Accurate dating remains critical to making inferences about timing and structure of reversals and excursions. Controversy remains about the significance of such features as the preferred longitudinal paths that virtual geomagnetic poles at some sites seem to follow during excursions and reversals. Reversal rates are estimated under the assumption that reversal occurrence times can be described as a Poisson process. Correlations are sought between reversal rates and other properties of the paleomagnetic secular variation, and more general models for reversals and secular variations are being developed to provide predictions of the power spectrum of geomagnetic intensity variations for comparison with those derived from long paleomagnetic records. These analyses may ultimately allow the identification of any characteristic timescales associated with the geomagnetic reversal process, and should prove useful in evaluating the behavior observed from numerical simulations of the geodynamo.

1. Introduction

In the early 1900s, study of the magnetization of rocks in France and Italy led Bernard Brunhes to conclude that Earth's magnetic field had formerly had the opposite direction to that observed today. The evidence for geomagnetic reversals was not, however, generally accepted as conclusive until the 1960s, by which time compelling evidence of global reversals was assembled for radiometrically dated rocks (Cox *et al.*, 1964). At about the same time Vine and Matthews (1963) interpreted the magnetic anomaly record from the Indian Ocean as the record in seafloor rocks of the reversing geomagnetic field. While it is plausible that the magnetic field has been reversing throughout most of its existence, the recovery of magnetostratigraphic records becomes increasingly challenging as the timescale is extended beyond the age of the seafloor. Prior to 2 Ga there is at present no record of a succession of several magnetic reversals, while between 2 and 1 Ga such records are rare (Pavlov and Gallet, 2001), and difficulties in obtaining precise dating constraints mean that even estimating the reversal rate is a far from straightforward task. Although the rate at which reversals occur is not uniformly well determined, it is nevertheless clear that the rate has varied throughout geological time and, perhaps more surprisingly, there exist a few long time intervals, when there have been very few or no reversals, the most recent example being the almost 40 Ma period known as the Cretaceous Normal Superchron, which ended at about 83 Ma.

Geomagnetic records in volcanic rocks and seafloor magnetic anomalies remain critically important in geomagnetic reversal studies today, although the emphasis has now shifted from establishing the global validity of geomagnetic reversals to understanding their detailed structure and origin and what this may reveal about the geodynamo (e.g. Jacobs, 1994; Merrill *et al.*, 1996). Sedimentary rocks can also preserve a record of the reversing geomagnetic field, although the timing of remanence acquisition and/or the possible smoothing of the geomagnetic record requires careful evaluation (e.g. Langereis *et al.*, 1992; Tauxe *et al.*, 1996; Clement, 2000). Typically, the paelomagnetic record preserved in volcanic and sedimentary rocks is used to study the details of directional and (where possible) intensity changes for transitions at individual locations, while the oceanic magnetic anomaly records provide the times of occurrence for reversals and thus reversal rate, although increasingly the marine record is being interpreted in more detail as attempts are made to separate the effects of paleointensity fluctuations from those attributable to variable mineral magnetism in the magnetic anomaly record (see, e.g. Kent and Carlut, 2000; Gee *et al.*, 2000; Pouliquen *et al.*, 2001).

Geomagnetic reversals remain a fascinating but rather poorly understood phenomenon, in part because of the fragmentary and inherently inaccurate nature of the geological record. From paleomagnetic records it is known that the intensity of the magnetic field decreases drastically during reversals, that the field structure is nonzonal, and reversals take of the order of thousands of years. However, despite several decades of reversal studies, the existence of consistent field structures during reversals is debated, and the characteristic timescales remain poorly defined. Also despite considerable advances in numerical simulations of the dynamics of geomagnetic field behavior, such numerical models cannot be operated in the parameter regime considered appropriate for Earth, and therefore only provide general guidelines about the behavior to be expected for self-sustaining dynamos. This chapter is concerned with the paleomagnetic record of geomagnetic reversals and its interpretation in the light of both phenomenological models for geomagnetic reversals and the increasingly prevalent numerical models of the geodynamo. We first attempt some working definitions of geomagnetic reversals, and their relation to excursions and what is sometimes called normal secular variation. We consider the timescales involved in reversal processes, including changes in reversal rates, how the timescales are measured and the current limitations on such measurements. Knowledge of the timescales involved is critical to the resolution of such controversial issues as whether the geomagnetic pole follows a prefered path or paths during reversal, or preferentially occupies particular geographic locations for extended periods of time.

We conclude with a look towards what may or may not be resolved in the near future by reversal studies, and the prospects on how they may inform us on the workings of the geodynamo.

2. Working definitions for reversals and excursions

When it is not in a transitional state the geomagnetic field can be represented to first order by that of a geocentric dipole. This representation has become an entrenched part of paleomagnetic analyses (even when the field is reversing), so that it is common for measurements of the magnetic field to be expressed in terms of virtual geomagnetic pole (VGP) positions and virtual dipole moments (VDMs). A VGP gives the geographic coordinates on Earth's surface of the North Polar axis of the geocentric dipole that would generate the observed local field direction, and effectively removes the gross geographic variations attributable to a field of dipolar origin. We think of the VGP as a unit vector $\hat{\mathbf{v}}$, associated with a vector dipole moment $\mathbf{V}$ in a geocentric coordinate system associated with a local field vector $\mathbf{B}$ at a site location at $\mathbf{s}$. Although the VGP transformation is conventionally given in terms of spherical trigonometry (e.g. Merrill *et al.*, 1996) it is in fact a linear transformation on $\mathbf{B}$ as indicated by Egbert (1992): that is, we can write

$$\mathbf{V}(\mathbf{s}) = R(s, \theta, \phi)\mathbf{B}(\mathbf{s})$$

with R at a location with radius s, colatitude θ, and longitude ϕ given explicitly by

$$R(s, \theta, \phi) = \frac{4\pi s^3}{\mu_0} \begin{pmatrix} -\cos\theta\cos\phi & \sin\phi & \frac{1}{2}\sin\theta\cos\theta \\ -\cos\theta\sin\phi & -\cos\phi & \frac{1}{2}\sin\theta\sin\theta \\ \sin\theta & 0 & \frac{1}{2}\cos\theta \end{pmatrix}$$

while a VDM, V, is the magnetic moment $|\mathbf{V}|$ that a geocentric dipole would have in order to generate the observed field intensity. Thus

$$V = |\mathbf{V}| \qquad \hat{\mathbf{v}} = \mathbf{V}/|\mathbf{V}|.$$

Although $\hat{\mathbf{v}}$ can be derived from directional measurements alone, V requires knowledge of the magnetic inclination in addition to field magnitude. A variant on these transformations is the virtual axial dipole moment (VADM) which is the dipole moment corresponding to a geocentric dipole aligned with the geographic axis. Although VGPs and VDMs are widely thought of as representations that approximate the geomagnetic field as a dipole it can be seen from the above that they are simply linear transformations of the local magnetic field vector, and still contain all the nondipole contributions to the magnetic field. They can thus be used to describe any geomagnetic field structure whether or not it is predominantly dipolar, but the interpretation of the nondipolar field contributions is less obvious than for a dominantly dipolar field. Figure 1(a) shows the concentration distribution of VGPs, derived from a uniform sampling of the geomagnetic field in 2000 (predicted from Ørsted initial field model (OIFM) of Olsen *et al.*, 2000). The VGPs are most concentrated about the geomagnetic dipole axis reflecting the predominantly dipolar structure, however, there are occasional outliers up to 30° or 40° from the dipole axis, reflecting the second- and higher-order structure of the field. A purely dipolar field would have all the VGPs concentrated at a single point corresponding to the dipole axis.

During a reversal the axial dipole part of the field is expected to be greatly diminished for a prolonged time interval and must certainly vanish at some point. In Figure 1(b), the axial dipole part of the field is removed from OIFM revealing a field that may be similar in some ways to what might be expected during a reversal, and the second order structure exhibits a different latitudinal distribution for $\hat{\mathbf{v}}$: there are particularly heavy concentrations just east of the Americas, and subsidiary concentrations in parts of the Pacific and in a longitudinal band extending from Eastern Asia down through Western Australia. These concentrations arise from the higher order structure in the field. The origin of these concentrations can be seen in Figure 2(a) where the actual latitudes for each VGP from OIFM are contoured for geographic locations. From this we can see which locations contribute the VGPs with greatest departures from the geographic axis: in (a) the lowest VGP latitudes for OIFM occur in the southern Indian Ocean and around Antarctica. The picture appears quite different when the axial-dipole part of the field is removed in (b), exposing VGP latitudes less than zero over

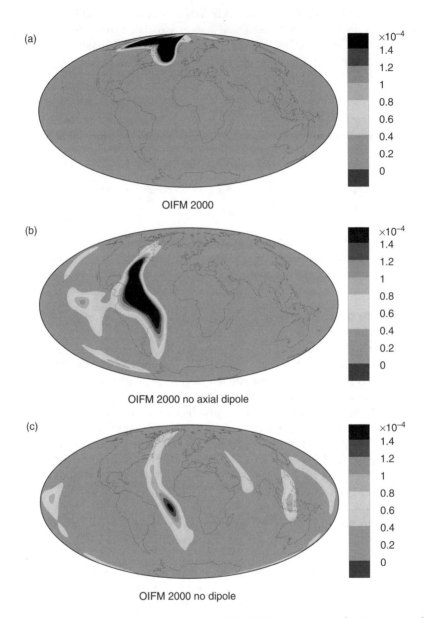

Figure 1 Probability density of VGP positions for OIFM 2000 (a) includes g_1^0, (b) omitting g_1^0, and (c) omitting all dipole terms. (See Colour Plate VIII.)

large regions in both the northern and southern hemisphere. Figure 3(a) and (b) shows the values of VGP longitudes as a function of geographic location: they are identical because g_1^0, the axial-dipole field contribution has no longitude dependence. The difficulty in rapidly comprehending the structure of the geomagnetic field from concentrations of $\hat{\mathbf{v}}$ or its location for individual sites can be readily appreciated from the differences between Figures 1–3 parts (b) and (c). In part (c) of each figure the entire dipole contribution, as opposed to just

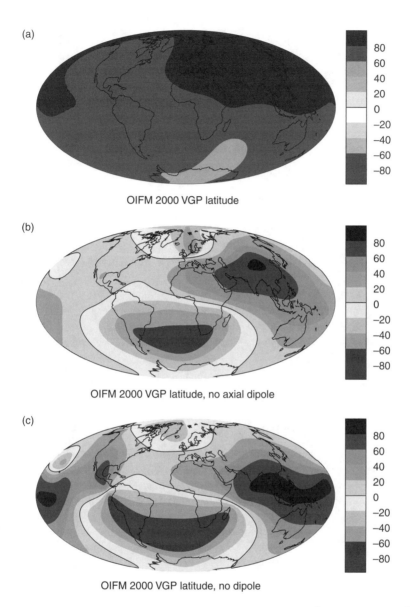

Figure 2 Predicted VGP latitudes from OIFM 2000 (a) including g_1^0, (b) omitting g_1^0, and (c) omitting all dipole terms. (See Colour Plate IX.)

the axial dipole in (b), has been removed from OIFM: it is seen that $\hat{\mathbf{v}}$ and its statistical distribution can look quite different depending on the form of the equatorial dipole terms once the axial dipole is removed or much reduced in magnitude. Since the equatorial dipole terms are known to change on relatively short timescales (e.g. Constable *et al.*, 2000), and their overall contributions to the dipole field may be quite small, there might be good reasons for considering Figures 1(c)–3(c) as at least as likely to be representative of what goes on during a reversal as Figures 1(b)–3(b).

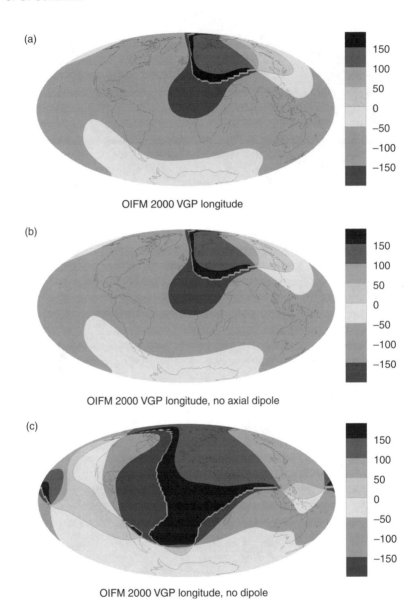

(a)

OIFM 2000 VGP longitude

(b)

OIFM 2000 VGP longitude, no axial dipole

(c)

OIFM 2000 VGP longitude, no dipole

Figure 3 Same as Figure 2, but for VGP longitude. Note that the VGP longitudes are the same whether or not the axial-dipole term is included, because it has no dependence on longitude. (See Colour Plate X.)

A time interval in which the geomagnetic field is predominantly of one polarity is known as a chron: typical lengths range from the order of 0.1–1 Ma, with those rare extremely long intervals dominated by one polarity being designated superchrons. Short intervals of stable polarity ∼10–100 ka are called subchrons. Transitional geomagnetic field directions are often defined in terms of a minimum angular departure of $\hat{v}$ from the rotation axis: the numerical value typical ranges from about 40° to 50°, and reflects the perception that the average dispersion about this axis during stable polarity times ranges from about 12° at the equator to somewhere around 20° near the pole (e.g. McFadden *et al.*, 1991). The root mean square value of this angular deviation of $\hat{v}$ from the rotation axis is widely interpreted as a measure of the variability in the geomagnetic field due to secular variation for nontransitional magnetic field states. Merrill *et al.* (1996) define a geomagnetic reversal as a "globally observed 180° change in the dipole field averaged over a few thousand years." The generality of this definition arises from the genuine difficulty in identifying from paleomagnetic records the time at which the field is irrevocably committed to reversing rather than returning to its starting polarity: it appears that only after the fact can one be sure that a geomagnetic reversal has taken place. Brief excursions of the geomagnetic field have been documented in which at some locality the VGP position is observed to deviate more than some threshold amount from the geographic axis (e.g. Verosub, 1977, specified an angular deviation of $> 45°$), and then return to the original polarity rather than executing a full reversal. Such excursions are usually accompanied by decreases in the local geomagnetic field intensity. Some of these excursions are documented in a number of locations with sufficient precision in age that they are taken to be contemporaneous (e.g. the Laschamp excursion occurring at about 40 ka, Chauvin *et al.*, 1989; Levi *et al.*, 1990; Thouveny *et al.*, 1992; Vlag *et al.*, 1996; Nowaczyk and Antonow, 1997). A widespread interpretation of excursions is as aborted reversals in which the field returns to its original state after a large perturbation. Observationally, excursions are distinguished from subchrons by the fact that the field does not apppear to achieve a fully reversed stable state. Features that could be thought of in the same category as excursions are the cryptochrons, events observed in magnetic anomaly data that typically last less than 30 ka (Cande and Kent, 1992): although the resolution of the magnetic anomaly record is only rarely sufficient to distinguish whether these correspond to subchrons, excursions, or geomagnetic intensity fluctuations.

Recent articles on dating short geomagnetic field events during the current polarity interval find 10 or more excursions of the geomagnetic field (Langereis *et al.*, 1997; Lund *et al.*, 1998) during the Brunhes chron. In comparison with the last reversal, which occurred at 0.78 Ma, these are rather poorly documented, with at most a handful of sites recording any individual event. In contrast, Love and Mazaud (1997) in constructing a database of records of the last reversal derived from sedimentary rocks or lava flows uncovered 62 distinct records of the Matuyama–Brunhes transition, although after applying their selection criteria concerning number of time samples and lab analyses they were only able to retain 11 of these for more detailed analyses of field structure. In the time interval 0–158 Ma, there are about 295 reversals in the marine magnetic anomaly record plus another 112 events identified as cryptochrons (Harland *et al.*, 1990; Cande and Kent, 1992; Constable, 2000). Because of the uneven temporal resolution in the record more short events may be identified in the future. For the vast majority of known reversals no detailed geological record has yet been uncovered.

Figure 1 shows that in terms of individual site records using departures of say 45° or more of VGPs from the geographic axis as the criterion for identifying excursional observations is inadequate. Although the present field is stable by this criterion [Figure 1(a)], with the axial dipole contribution to the present field removed as in (b) there are places with VGP latitudes close to 90°, which by a casual interpretation would not appear to be taking part in any excursional activity, despite the absence of a dominant axial dipole. The same holds true, although to a lesser degree, in (c). Yet by any reasonable criterion the fields in both (b) and (c) would be considered anomalous when compared with the present field.

A plausible interpretation of what can be found in the paleomagnetic record is that the geomagnetic field exhibits a continuum of behavior in its secular variation so that in fact there is no clearly defined excursional state, but occasional large departures from the average state which may or may not ultimately result in a reversal. This idea ties in quite well with other views about the importance of the inner core in stabilizing numerical and by inference real dynamos to produce more Earth-like behavior (Hollerbach and Jones, 1993a,b, 1995; Clement and Stixrude, 1995; Gilder and Glen, 1998). It is also consistent with an absence of identifiable triggers or other signature features for the reversals that are seen in numerical simulations of geodynamos (e.g. Coe *et al.*, 2000; McMillan *et al.*, 2001). Gubbins (1999) has suggested that a possible distinction between excursions and reversals may be that during excursions the field may reverse in the liquid outer core, which has typical timescales of 500 years or less, but not in the solid inner core where the relevant diffusion timescale is several thousand years. In order for a reversal to be carried through to completion the reverse polarity must diffuse into the inner core. One possible implication is that reversals and excursions could have different characteristic timescales in the associated magnetic field changes. This highlights the importance of being able to determine the age and duration of geomagnetic events of all types, a topic to be discussed in the next section.

3. Dating and timescales

Dating for paleomagnetic records, and the geomagnetic information that comes with it, comes from a number of sources with assorted levels of accuracy and resolution and corresponding to applications in studying particular geomagnetic phenomena. The crudest form of geomagnetic information (the polarity of the field) comes in the form of the global magnetostratigraphic timescale, which supplies a time at which the geomagnetic field reversed, with the associated implication that at this level of resolution it can be considered instantaneous. The absolute timing of the magnetic anomalies is linked to age provided for specific isotopic tie points in the scale (see e.g. Cande and Kent, 1995). The accuracy of the timescale depends on the accuracy of the tie points and the validity of the assumptions concerning constancy of seafloor spreading rate between tie points. The resolution provided by picking anomalies ranges from about 10 to 50 ka depending on the spreading rate and water depth for the particular marine survey. The Cande and Kent (1995) timescale, for example, aims to have a uniform resolution in which all intervals longer than 30 ka are included. Short events may be unresolvable in this kind of record, as will excursions and any transitional states that may exist during reversals. Evidence for the existence of such features is provided by the cryptochrons, finer scale features documented by Cande and Kent in the marine magnetic anomaly record back to 83 Ma. Some of these cryptochrons correspond to short events of stable magnetic polarity, while others may reflect changes in the paleointensity of the field. A distinction between these two causes cannot be made solely on the basis of the magnetic anomaly record: other paleomagnetic data are required. The magnetic anomaly timescale only extends to about 160 Ma because of the scarcity of older seafloor.

An alternative method is used for finding the age of an individual reversal boundary for land-based paleomagnetic measurements and requires knowledge of the magnetic polarity and a radiometric date for a collection of different rock samples. In the Chronogram technique (Cox and Dalrymple, 1967) one finds the most probable age for individual reversal boundaries, by minimizing weighted squared deviations of a collection of K/Ar or Ar/Ar ages for a range of test boundaries. Limitations of this method (Tauxe *et al.*, 1992) are that one can only deal with data surrounding one reversal at time, the presumption that the polarity can be determined unambiguously, nonuniformity in the distribution of dates, and potentially inadequate estimates of the uncertainties in dating.

High-resolution records of geomagnetic field behavior are now routinely recovered from marine sediments, and these can be dated using the so-called astronomical timescale (ATS). ATS makes use of variations in Earth's orbital parameters (precession, obliquity, and eccentricity) for last few million years. The climatic response to these variations of ice sheet growth and decay can be approximately modelled. Then if there exists a high-resolution record of Earth's response to this orbital forcing that can be tied into magnetostratigraphy this can be used to provide dates for reversals. Such a high resolution record was first provided in the form of fluctuations in the oxygen isotopes, $\delta^{18}O$. Johnson (1982) found a result from this method of 0.79 Ma, which conflicted with the 0.73 Ma age for the Brunhes–Matuyama boundary that was widely accepted at the time. This conflict was initially attributed to insufficient high-resolution records or uncertainties concerning the amount of time taken for magnetization to be locked into the continuously deposited sediment. In a later study, Shackleton *et al.* (1990) used a detailed $\delta^{18}O$ record from Ocean Drilling Project Site 677 to place the Brunhes–Matuyama boundary at 0.78 Ma. A study from the Mediterannean by Hilgen (1991) on fluctuations in sapropels also associated with climate variations produced the same result. These results, combined with new radiometric measurements prompted a re-examination of the reliability of the 0.73 Ma age derived from radiometric dating. Tauxe *et al.* (1992) showed using a simple bootstrap resampling scheme that the error bars on reversal times derived by the chronogram technique may be unrealistically small because of violations of assumptions made in the method, and thus that radiometric ages for the Brunhes–Matuyama boundary were not inconsistent with those derived via the ATS.

In the past decade the times at which the most recent reversals occurred have been estimated by direct $^{40}Ar/^{39}Ar$ dating of lava flows that record transitional paleomagnetic directions (e.g. Baksi *et al.*, 1992; Singer and Pringle, 1996; Singer *et al.*, 1999). The dating techniques now in use have a claimed precision of 0.5% on quaternary sanidine, or 0.5–1% on basaltic or andesitic lavas. These results confirm that the age of the most recent reversal is 0.78 Ma, and have spawned a new approach to calibrating the most recent part of the magnetostratigraphic timescale, and studying the structure of the geomagnetic field during reversals. This will be discussed in more detail in the Section 5 on "Structure and duration of reversals."

4. Reversal rates

4.1. Measuring rates

The fundamental question that can be addressed using the information provided by the magnetostratigraphic timescale concerns the estimation of the frequency with which geomagnetic reversals recur. Since the earliest studies on this topic, it has been apparent that the time elapsed between successive field transitions is highly variable. In the absence of a clear physical understanding of any fundamental mechanism and associated dynamics, reversals have been modeled in a phenomenological way. They have (following Cox, 1968) been regarded as uncorrelated, triggered events described by a Poisson process with a varying rate parameter. The rate parameter provides a statistical characterization for their times of occurrence, but it should be kept in mind that there is no direct physical model that associates it with equations describing the physical evolution of the magnetic field in Earth's core. Figure 4 shows estimates of geomagnetic reversal rate as a function of time for the last 160 Ma as determined by Constable (2000). The solid black line is based on adaptive

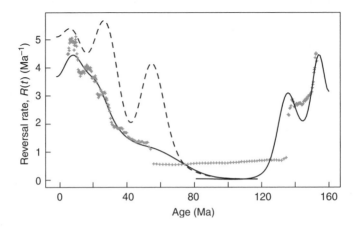

Figure 4 Reversal rates for past 160 Ma derived by Constable (2000) using kernel density estimates. Dashed line is rate including cryptochrons, solid line without. Light gray symbols give rate for sliding window estimate.

Gaussian kernel estimates, in which each reversal is given unit weight, and the Cande and Kent (1995) timescale is used for 0–84 Ma, and Harland *et al.* (1990) prior to 84 Ma. Gray symbols are 50-point sliding window estimates, which cannot be considered independent, because of the overlap of adjacent windows. The dashed black line gives the kernel estimate when events identified by Cande and Kent as cryptochrons are treated as reversals. Whether these are very short polarity intervals, intensity fluctuations, or (perhaps more likely) some of both remains an open question. It is seen from this figure that there are long-term changes in reversal rate: these have been variously attributed to gradual thermal changes in conditions at the core–mantle boundary (CMB) (McFadden and Merrill, 1984), the release of mantle plumes (Loper, 1992; Loper and McCartney, 1986; Courtillot and Besse, 1987; Larsen and Olson, 1991) or the sudden arrival of cold material at the CMB from a mantle flushing event (Gallet and Hulot, 1997). A possibility that cannot be unequivocally ruled out, is that the apparent long-term trend in reversal rates is simply what can be expected from a chaotic process producing the magnetic field in Earth's core.

   The degree to which finer scale structure in the reversal rate curve should be interpreted as reflecting changes in the geodynamo has been hotly debated. Aside from the obvious limitations of such a statistical description, difficulties in estimating the rate parameter from small numbers of events mean that short-term fluctuations in the rate cannot be reliably estimated. McFadden and Merrill (1984) considered the reversal rate, $\lambda(t)$, to be varying in a piecewise linear fashion with time, with zero rate during the Cretaceous Normal Superchron (CNS). Rather than assuming a specific functional form for the temporal variation, Constable (2000) used a technique proposed by Hengartner and Stark (1992a,b, 1995) to compute pointwise confidence bounds on the temporal probability density function for geomagnetic reversals. The method allows the computation of a lower bound on the number of modes required by the observations, thus enabling a test of whether "bumps" are required features of the reversal rate function. Conservative 95% confidence intervals can then be calculated for the temporal location of a single mode or antimode of the probability density function. Using observations from the time interval 0–158 Ma it is found that the derivative

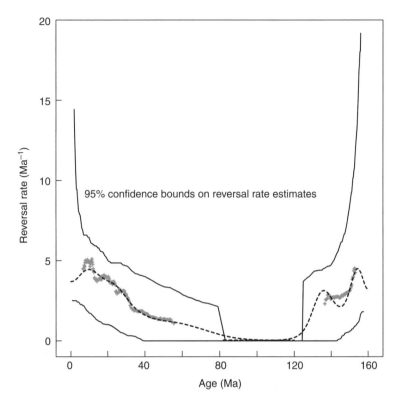

Figure 5 Reversal rates, excluding cryptochrons as in Figure 4, dashed line is kernel estimate, symbols sliding window rate, and solid lines are 95% confidence bounds on reversal rate under assumption that reversals are produced by a Poisson process with monotonically decreasing rate parameter from 160 to 124 Ma, zero rate during the CNS, and montonically increasing from 83 Ma to present.

of the rate function must have changed sign at least once. The timing of this sign change is constrained to be between 152.56 and 22.46 Ma at the 95% confidence level. Confidence bounds are shown in Figure 5 for the reversal rate under the assumption that the observed reversals are a realization of an inhomogenous Poisson or other renewal process with an arbitrary monotonically increasing rate function from the end of the CNS to the present, a zero rate during the CNS, and a monotonically decreasing rate function from M29R at 158 Ma to the onset of the CNS. The confidence limits on estimating the rate parameter are broad in comparison to the size of the shorter term fluctuations in rate. It is unnecessary to invoke more than one sign change in the derivative of the rate function to fit the observations.

There is no incompatibility between these results and a recent assertion that there is an asymmetry in average reversal rate prior to and after the CNS (McFadden and Merrill, 1997), when the CNS is assumed to be a period of zero reversal rate. Neither can the Poisson model be used to reject an alternative hypothesis that rates are essentially constant from 158 to 130 Ma, and from 25 Ma to the present, with an intermediate nonstationary segment as envisaged by Gallet and Hulot (1997).

4.2. Reversal rates and other properties of the geomagnetic field

The rate parameter alone is unlikely to provide an understanding of what controls changes in reversal rate, and we must turn to other kinds of paleomagnetic observations and numerical simulations to provide a guide in determining how rapidly it is possible for the rate to change, and what controls such changes. To date, no long-term correlation has been found between average absolute intensity of the paleomagnetic field and long-term reversal rate (Selkin and Tauxe, 2000), although one was suspected for some time (see Prévot *et al.*, 1990). It should be emphasized, however, that there remain insufficient data to characterize absolute paleointensity very well over the past 300 Ma, and in the CNS a period which is of great interest because of the absence of documented reversals there is only a handful of reliable absolute paleointensity data available despite continuing efforts to acquire such observations (see e.g. Tarduno *et al.*, 2001). The difficulty in acquiring reliable data is in large part due to the scarcity of suitable geological formations with lava flows in the right age interval.

The general paucity of absolute intensity data is to some extent mitigated by the increasing number of paleomagnetic records being acquired from marine sediments. These can provide long, almost continuous, records of both relative paleointensity and directional variations, although it is often the case that only one or the other is acquired. Such records are highly variable in both duration and resolution, but in the context of reversal rates it is worth noting that there are now at least three records of relative geomagnetic intensity available each spanning 4 million or more years of field variation at a single location (Valet and Meynadier, 1993; Constable *et al.*, 1998; Cronin *et al.*, 2001).

Constable *et al.* (1998) did find in their analysis of an 11 Ma long record from a single site (ODP 522, with paleolatitude 33°S during the Oligocene that there was a weak correlation between relative paleointensity and polarity interval length. They also found differences in the power spectra of the intensity record. In the time interval 23–28 Ma, where the reversal rate is $4\,\mathrm{Ma}^{-1}$, the spectrum of intensity variations is dominated by the reversal process in the frequency range 1–$50\,\mathrm{Ma}^{-1}$. In contrast, between 34.7 and 29.4 Ma, when the rate is about $1.6\,\mathrm{Ma}^{-1}$, the field is stronger, more variable, and a strong peak occurs in the spectrum at about $8\,\mathrm{Ma}^{-1}$ (see Figure 6). Constable *et al.* (1998) speculate that this peak is associated with the cryptochrons found in this time interval, although the time resolution in the record is not adequate to provide a definitive correlation with the magnetic anomaly record.

Analysis of Valet and Meynadier's (1993) 0–4 Ma paleointensity record from ODP sites 848, 851, and 852 in the equatorial Pacific reveals a spectral structure that is similar to that from ODP 522 in being dominated by power at low frequencies. When the errors in spectral estimation are taken into account it is unlikely that the spectra can be considered significantly different. However, the rapid drop in power between 0 and $5\,\mathrm{Ma}^{-1}$ is distributed to slightly higher frequency in the more recent record, reflecting the influence of the sawtooth signal noted by those authors. The overall power in relative intensity variations is quite similar.

In a paleosecular variation study using Cretaceous age marine sediments collected in the Umbria-Marche region of Italy, Cronin *et al.* (2001) derived detailed relative paleointensity and direction records for a 4 Ma segment of the CNS extending from about 90 to 94 Ma, and find no convincing evidence for excursions. The Umbrian relative paleointensity record indicates a subdued variation in paleointensity during the CNS when paleointensity is normalized relative to the mean within each record. During the Oligocene when the field reverses, the standard deviation in intensity is about 50% of the mean value (for both low and high reversal rate periods), as it is during the past few million years (Valet and Meynadier, 1993), while for the CNS data it is reduced to about 28% of the mean. The normalized intensity data

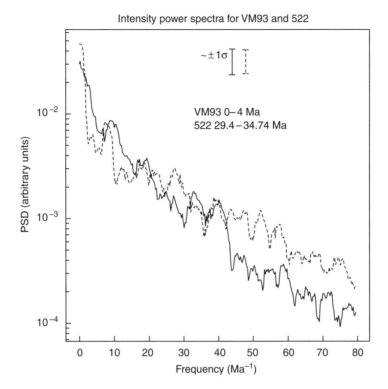

Figure 6 Power spectra derived from relative geomagnetic paleointensity records for 0–4 Ma (Valet and Meynadier, 1993), and low reversal rate time intervals during the Oligocene (Constable *et al.*, 1998). Larger error bar is for the VM93 spectral estimate, dashed line is 522 and solid line is for VM93 record.

distributions are clearly different, a fact that we may attribute to the difference in reversal rates between the two time intervals.

One might also expect greater directional stability for the magnetic field during times of low reversal rate. McFadden *et al.* (1991) show that the dispersion in VGP directions derived from globally distributed lava flows in the time interval 80–110 Ma is lower than during 0–5 Ma and have argued that the contribution of symmetric family spherical harmonics to VGP dispersion is decreased during the CNS. Cronin *et al.* (2001) find a very similar level of directional stability to that found by McFadden *et al.* for the time period 80–110 Ma for lava flows, despite the large differences in the nature of the two data sets: the lava flows are globally distributed spot readings of the field distributed over a 30 Ma long interval which includes the transition to a period when reversals begin to occur again; in contrast each sediment datum from the ~4 Ma interval provides an average over about 3,000 years at a single site whose paleolatitude is inferred to be 21°. There are also large differences in the size of the two data sets: Cronin *et al.* acquired over 800 measurements of direction and intensity; the lava flows provide a few hundred estimates of direction globally, but only 18 within the same latitude band (10–25°), although they are individually considerably better determined than the sedimentary observations. Nevertheless, the combined evidence from these two kinds of observations point to decreased variance in $\hat{\mathbf{v}}$ during the CNS. Unfortunately no directional data exist for the Oligocene record which would allow such a comparison between reversals occurring at low and relatively high rates for that portion of the timescale.

4.3. *Longer-term rate variations, numerical dynamo simulations, and boundary conditions*

The above discussion of reversal rate data has been limited to the time interval covered by the marine magnetic anomaly record for which a more or less complete record of geomagnetic reversals is available. There is evidence for large rate changes further back in time, and for at least one additional long interval of very low reversal rate the Permo-Carboniferous Reversed Superchron (PCRS) from about 302 to 252 Ma. Johnson *et al.* (1995) have used indirect statistical arguments, with a method proposed by McElhinny (1971) and selected observations from the Global Paleomagnetic Database (Lock and McElhinny, 1991; McElhinny and Lock, 1993) to construct a model of relative reversal rate based on the fraction of mixed polarity results in moving windows of length 20 Ma for the past 570 Ma. Their analysis suggested the possibility of a further time interval around 500 Ma ago when the reversal rate was much reduced, and the polarity predominantly reversed. Magnetostratigraphic work by Pavlov and Gallet (2001) confirms that a period with few or no reversals exists during the Ordovician, although its exact extent and the rates in surrounding intervals are not yet completely defined. The approximate time interval of around 150–200 Ma between successive superchrons is often equated with typical timescales for the overturn of the mantle, and thus linked to changes in CMB conditions.

Investigations of both kinematic dynamos and full scale numerical simulations have been used to demonstrate the potential importance of boundary conditions in controlling geomagnetic reversals. Hollerbach and Jones (1993a,b, 1995) demonstrated that a finitely conducting inner core played a critical role in stabilizing the magnetic field between reversals. It is thus plausible to suppose that growth of the inner core may in the past have had significant influence on the average structure of the geomagnetic field and dynamics in the outer core. Kent and Smethurst (1998) have suggested that pre-Mesozoic (older than 250 Ma) paleomagnetic data can be interpreted as offering some suppport for this kind of scenario: the abundance of shallow inclinations are compatible with substantially larger octupole contributions (~25% of the axial dipole) to the field than seen in more recent times. However, note Tauxe's (1999) caution that such pre-Mesozoic data may be biased by noise contributed from inadequate knowledge of paleohorizontal. Kent and Smethurst attribute the additional field complexity to a smaller size for the inner core at that time, although Bloxham (2000b) found that in his numerical dynamo simulations reasonable variations in the size of the inner core (for the relevant time interval) had no detectable influence in increasing the complexity of the average field. While it is conceivable that the size and geometry of the outer core influence the propensity of the geomagnetic field to reverse, this seems unlikely to be a major contributor to the variations in reversal rate in the Cenozoic, because at this point in Earth's history the inner core is presumably only growing very slowly.

Temporally changing and laterally varying CMB conditions are viable candidates for changes in the field structure and reversal rate. The influence of various distributions of heat flux at the CMB have been explored in numerical simulations (Glatzmaier *et al.*, 1999; Coe *et al.*, 2000; Bloxham, 2000a,b; see Dormy *et al.*, 2000 for a review). Bloxham concludes that Y_2^0 thermal structure can contribute to nonaxial-dipole field structure, while Glatzmaier *et al.* infer that the propensity for the field to reverse is greatly enhanced by particular kinds of boundary conditions, such as the one that mimics the current thermal structure near the CMB.

Quite different zonal thermal structures appear to be associated with simulations that do not exhibit reversals. It is difficult to envisage being able to run the current full three-dimensional numerical dynamos for long enough to generate adequate numbers of reversals to acquire statistically meaningful estimates of rate for a variety of boundary conditions, however, one can readily evaluate whether the spectrum of intensity variations (for example) is different, and explore whether it is possible to identify differences in the dynamical processes in the core associated with the different boundary conditions. It may be possible to make progress in understanding dynamical regimes that favor reversals, or why in some parts of the timescale cryptochrons appear quite common, while in others they are apparently absent. McMillan *et al.* (2001) have developed statistical tools for studying the output of numerical dynamo simulations and comparing them to properties of the paleomagnetic field. These have been applied to some of the Glatzmaier–Roberts numerical experiments with different boundary conditions, and clearly demonstrate more power in the secular variations for some scenarios than others.

An outstanding question, which has perhaps been obscured by the repeated analysis of paleomagnetic observations in terms of the Poisson process is how the geomagnetic field evolves from a state in which geomagnetic reversals occur to one without (or vice versa). Rapid changes in the rate parameter of a stochastic renewal process cannot be estimated accurately: yet it seems that the geomagnetic field readily evolved from a time when no reversals occurred (CNS) to one of quite frequent reversals over a time interval of some tens of millions of years. Such a timescale is significantly shorter than that generally attributed to changes at the CMB due to overturn of the mantle, so that one either has to invoke gradual changes that eventually pass some critical point allowing the initiation of reversal (as envisaged by, e.g. McFadden and Merrill, 2000), or alternatively one needs to investigate scenarios that involve mantle plumes or other processes in which significant changes in CMB conditions can be generated relatively rapidly.

5. Structure and duration of reversals

We turn now to the details of geomagnetic reversals, as opposed to the rate at which they occur. Once again there needs to be a critical emphasis on dating of paleomagnetic records, because this determines our ability to study such issues as how long the field spends in transition at any given site during any particular reversal. It is already obvious from Figure 1 that the duration of a reversal is likely to depend on geographic location if we use any of the standard measures to define a reversal, such as local VGP direction deviating from the geographic axis by more than a specified amount or decrease in paleointensity (DIP) to less than some critical percentage of the average. Note that this requires a clear concept of what the average is, something that is in general far from obvious.

5.1. *Observational estimates of duration*

There are four potential sources of observational estimates for the duration of reversals, each related to the nature of the geological features preserving the record. We first consider the direct estimates. In marine sediments, the sediment accummulation rate (SAR) can be used to find the age at which the transition begins and is completed (as defined by some specified directional deviation or DIP). This suffers from a number of disadvantages: the interval length is subject to fluctuations in SAR, to whatever interpolation used, and to smoothing or offsets in timing due to the lock in depth and filtering process (which may themselves depend on SAR) in the acquisition of remanence. These limitations are offset by the significant advantage that direct correlation to the ATS is possible in many cases using fluctuations in δ^{18}O. The length of reversals found by this technique range from as short as 100 years (Worm, 1997) to more than 20 ka (Herrero-Bervera *et al.*, 1987): it is very difficult to sort out the roles of the various individual factors that contribute to the large range, but in general shorter estimates are found when only directional information is considered than when preliminary decreases in paleointensity are taken into account. In a study of eight Pacific marine cores Clement and Kent (1984) found indications that reversals took about twice as long at mid-latitudes as at low latitudes, but Valet *et al.* (1989) found no such relationship for four Atlantic cores from the latitude range 0–50°N. More recently, in a detailed study of sedimentary records of the Matuyama–Brunhes reversal, Tauxe *et al.* (1996) found tantalizing suggestions that the age at which it occurred depended on geographic location, but were unable to rule out the possibility of variations in the timing at which the magnetic remanence is ultimately locked in.

The other direct approach uses radiometric dating on transitional lava flows from different locations, preferably potassium-rich sanidine bearing rhyolites. This has to date meant that the reversal is defined in terms of directional rather than intensity anomalies. As noted earlier, the claimed precision on ^{40}Ar/^{39}Ar dating may be as good as 0.5%: for the Brunhes this means relative uncertainities at different sites of a few thousand years. The differences among ages of eight transitional lava flows drawn from globally distributed locations (Chile, La Palma, Tahiti, and Maui) range up to 12 ka, placing this as a plausible upper bound on the duration of directional variations for the last reversal (Singer and Pringle, 1996). This is a global estimate for the duration of the reversal: it should be kept in mind that at any given location the transition might appear to be initiated at a later or earlier time and have occurred considerably more (or perhaps even less) rapidly.

The number of transitions that are dated directly in multiple locations is very small indeed. An indirect statistical approach that does not necessarily require precise radiometric dates on lava flows takes the ratio of the number of transitional to nontransitional directions in some time window, and supposes that this indicates the amount of time the field spends in transition. Such a method is inevitably complicated by excursions, although if one takes the view that they are part of the same process as reversals this may not matter. Nevertheless, it is worth noting that direct radiometric dating by Singer *et al.* (1999) finds at least seven and perhaps as many as 11 excursional directions, that could be interpreted as reversals attempts, in the time interval 1.18–0.78 Ma. Iceland is the archetypical location for duration estimates by such a statistical analysis as there are hundreds of flows that span several million years. Since the eruption of these flows can safely be regarded as uncorrelated with geomagnetic field behavior this can be presumed to provide an unbiased estimate of the time the field spends in whatever one defines as transitional directions. The value so obtained for Iceland is 6 ka.

Durations have also been estimated indirectly from the cooling rates of intrusives rocks. This method has only rarely been used and requires assumptions about the cooling process that are hard to verify, therefore we will not consider it further here. Of the above the first two provide the most direct estimates, but all may be potentially complicated by precursory excursions extending the length of reversal. It is also possible that any such excursional features may themselves be inextricably bound up with variations in field intensity.

The vast range of times estimated for various reversals at different times and places indicate the inadequacy of the current data set to resolve such issues as how long an individual reversal takes, whether the time taken varies with location, and/or with different reversals, and also whether it varies along with reversal rate. Although dating techniques are steadily improving, it seems unlikely that it will be possible to know ages to an accuracy much better than a few thousand years. Thus the possibility of directly modeling the structure of the geomagnetic field during a reversal remains remote (although this has not stopped people from trying, Mazaud, 1995; Shao *et al.*, 1999). There are no such dating limitations in the records of geomagnetic reversals provided by numerical geodynamo simulations. A detailed analysis of the geomagnetic reversals found in the Glatzmaier–Roberts dynamo (Coe *et al.*, 2000) indicates that reversal times vary widely depending on criteria used to determine onset of the reversal, but also with location, for successive reversals, and may exhibit different behavior depending on the thermal boundary conditions.

5.2. Persistent patterns?

Despite the poor prospect for recovering details of the geomagnetic field structure during geomagnetic reversals, there have been longstanding efforts to characterize general properties associated with geomagnetic reversals. There are a number of fairly recent reviews on the subject of reversals, which discuss in some detail the magnetization of rocks and attempts to categorize the morphology of geomagnetic reversals (see e.g. Jacobs, 1994; Merrill and McFadden, 1999). These morphological descriptions are generally known as phenomenological models, and have evolved from those proposed by Creer and Ispir (1970) in which the dipole part of the magnetic field either decayed away and subsequently rebuilt in the opposite direction or rotated without variation in dipole moment.

Observations that are universally agreed upon for geomagnetic field transitions are, first, that they are incompatible with a zonal magnetic field structure during reversals: if such a field were present then the VGP positions at all sites would follow a path along either the site longitude or its antipode: that this is not so has been apparent since work by Hillhouse and Cox (1976). The second feature common to all transition records is a significant decrease in field intensity: this appears to be a necessary but not sufficient condition, since records of paleointensity for the Brunhes (e.g. Guyodo and Valet, 1999) show many such decreases to about 20% of the average value: although a number of these may coincide with directional anomalies or excursions, only the Brunhes–Matuyama boundary actually has a full geomagnetic reversal.

On the subject of more detailed, persistent structure in the magnetic field during reversals, there has been considerable controversy regarding interpretation of the available data. Valet and Laj (1984) noted that three out of four reversals from Crete had longitudinally confined VGP paths, suggesting recurring behavior during successive reversals for a single site. Clement (1991) analyzed VGP positions for the Matuyama–Brunhes transitions from globally distributed sites and noted preferential longitudes for VGP paths in the available records: he suggested that there was a simple field structure during this reversal. Laj *et al.* (1991) asserted that the simple structure had persisted for many reversals. Both of these assertions were criticized on the grounds that the global distribution of data was inadequate, and the possibility that biases in acquisition of magnetization by marine sediments, and/or statistical biases arising from possible distributions of directions would tend to favor reversal paths lying 90° from the sites longitude, as was observed in a number of cases (e.g. Egbert, 1992; Langereis *et al.*, 1992: Valet *et al.*, 1992, plus many more). Constable (1992) noted that the preferred longitude bands were roughly aligned with one of those followed by globally distributed VGPs when the axial dipole part of the present field decays to zero amplitude before regrowing with the opposite polarity. Figure 1(b) shows the probability density function for VGPs at the midpoint of such a process when there is no axial dipole

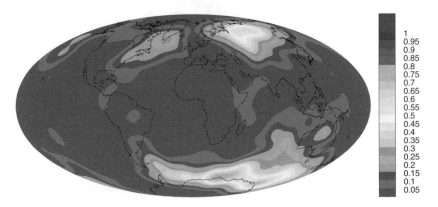

Figure 7 VGP density distribution for MBD97, the Matuyama–Brunhes transitional database compilation of Love and Mazaud (1997). Each direction is assigned unit weight in estimating the density function. (See Colour Plate XI.)

present. The coincidence between preferred paths and the present field could be interpreted as due to a permanently preferred field structure like that of the present day field or residual overprinting of paleomagnetic records by the present field.

The available reversal data have been reanalyzed a number of times with conclusion in favor of or against the preferred paths. Recently, 62 records of the Matuyama–Brunhes reversal were assessed by Love and Mazaud (1997), and winnowed to 11 in the MBD97 database. The remaining 11 still show preferred bands, even when weighted so that each site carries unit weight. Hoffman (2000) prefered to give each VGP direction (of which each record has a variable number) unit weight, and argues that when a running average of number of VGPs within circular caps is computed in this way, there is a preference for transitional VGP directions to lie in one of two preferred patches, north of Western Australia, and near the southern tip of America (see Figure 7, for the results from a similar calculation in which the density distribution for individual VGPs during the last reversal is calculated). He interprets this preference as a sign that the geomagnetic field occupied an intermediate dipolar structure with a pole in one of these positions for a significant amount of time during the reversal, while other parts of the transition were executed quite rapidly. In the absence of sufficiently accurate age constraints on the individual flows concerned, it is difficult to test this assertion. The prefered paths for transitional fields have been associated with thermal structure at the CMB (Laj *et al.*, 1991), as inferred from seismic tomographic models: difficulties with this interpretation remain in that it is not obvious what, if any, relationship should exist between the VGP positions and geomagnetic field structure and how either of these would be influenced by structure at the CMB.

Love (1998) also reanalyzed volcanic records of reversals and excursions for the past 20 Ma, and concluded that they too exhibit preferential paths in contradiction of a previous analysis by Prévot and Camps (1993). The major difference in the analysis was again in weighting, with regard to area dependence with latitude and number of intermediate directions for each record. The most significant contributors to the preferential longitudes were Icelandic records. Note the coincidence of this longitude with the VGP longitudes for Iceland for the present field seen in Figure 3. The question of why the present field should be of special significance during reversals is obscured somewhat by paleomagnetic evidence that the position of the dipole axis has been quite variable during the past few thousand years, and seems to show little preference for specific longitude bands (e.g. Constable *et al.*, 2000).

The discussion of persistent patterns remains inconclusive, as with other features, however, we can turn to numerical gedynamo simulations in an attempt to guide our understanding. Glatzmaier *et al.* (1999) have run numerical simulations of the geodynamo using a variety of thermal boundary conditions, some of which appear to be more susceptible to reversal than others, although the time series are not sufficiently long to gather reliable statistics on the matter. In their analyses of the homogenous and tomographic versions of the model both Coe *et al.* (2000) and McMillan *et al.* (2001) found that all reversals appear spontaneous, with no external triggering required. The tomographic model was the least stable. Coe *et al.* observed that the time taken for individual reversals ranges from 2–7 ka except for the second reversal in the tomographic model which takes 22 ka for the directional changes. The duration depends on the location, the reversal, and the criteria used to define transitional. All reversals were characterized by low intensities, and a sharp drop in dipole energy. The tomographic model of Glatzmaier *et al.* (1999) appears to offer some support for the hypothesis that VGPs during reversals correlate with areas of higher than average heat flux.

6. Future prospects in geomagnetic reversal studies

In concluding this review, it seems relevant to consider what prospects there are for improved understanding of geomagnetic reversals in the foreseeable future. In the paleomagnetic arena, understanding of timescales is likely to see a steady improvement, with more detailed records of $\delta^{18}O$ in marine sediments and a better understanding of how magnetic remanence is acquired (see e.g. Katari and Tauxe, 2000; Katari *et al.*, 2000; Katari and Bloxham, 2001). Use of anisotropy of magnetic susceptibility and increasingly sophisticated tests for both physical disturbance and rock magnetic changes in sediments (e.g. Cronin *et al.*, 2001) will allow better discrimination of which excursional records are reliable, as well as pinning down the timing of the multiple directional excursions that seem to occur in the late Matuyama and early Brunhes chrons. However, it remains unlikely that records with sufficient resolution and accuracy in dating will be acquired for a viable reconstruction of the geomagnetic field during a reversal.

On the other hand, there are real possibilities for improving understanding of the relationship between geomagnetic reversal rate and other properties of the geomagnetic field. At present there are three fairly detailed records of relative geomagnetic paleointensity 4 Ma long or longer (from 0 to 4 Ma, Valet and Meynadier, 1993, the Oligocene, 23–35 Ma and the CNS, approximately 90–94 Ma): it is possible to see the beginnings of associating reversal rates with particular kinds of geomagnetic field behavior. In the Oligocene, low reversal rates correspond to higher average field intensity. Such high intensities seem to be associated with strong geomagentic field variations. During the Cretaceous we do not yet have a clear view of absolute paleointensity, but the field shows a remarkable degree of stability as measured by the variability relative to the mean. Additional absolute paleointensity measurements for the CNS are in the pipeline (Lisa Tauxe, 2001, personal communication; Tarduno *et al.*, 2001). An increasingly detailed view of relative paleointensity variations is being acquired for the Brunhes (Guyodo and Valet, 1999) with the advent of the SINT800 model, and numerical geodynamo simulations allow the investigation of the limitations of stacking of paleomagnetic records from diverse locations. It seems likely that this progress in data acquisition will allow us to move on from the primitive view of reversals as symptoms of a Poisson process at least to a statistical description with arbitrary temporal correlations.

References

Baksi, A. K., Hsu, V., McWilliams, M. O., and Farrar, E. "$^{40}Ar/^{39}Ar$ dating of the Brunhes–Matuyama geomagnetic reversal," *Science* **256**, 356–357 (1992).

Bloxham, J., "The effect of thermal core–mantle interactions on the palaeomagnetic secular variation," *Phil. Trans. R. Soc. Lond.* A**358**, 1171–1179 (2000a).

Bloxham, J., "Sensitivity of the geomagnetic axial dipole to thermal core–mantle interactions," *Nature* **405**, 63–65 (2000b).

Cande, S. C. and Kent, D. V., "Ultrahigh resolution marine magnetic anomaly profiles: a record of continuous paleointensity variations," *J. Geophys. Res.* **97**, 15075–15083 (1992).

Cande, S. C. and Kent, D. V., "Revised calibration of the geomagnetic polarity timescale for the late Cretaceous and Cenozoic," *J. Geophys. Res.* **100**, 6093–6095 (1995).

Chauvin, A., Duncan, R. A., Bonhommet, N. and Levi, I. S., "Paleointensity of the Earth's magnetic field and K–Ar dating of the Louchadière volcanic flow (Central France) – new evidence for the Laschamp excursion," *Geophys. Res. Lett.* **16**, 1189–1192 (1989).

Clement, B. M., "Geographical distribution of transitional VGPs: evidence for non zonal equatorial symmetry during the Matuyama–Brunhes geomagnetic reversal," *Earth Planet. Sci. Lett.* **104**, 48–58 (1991).

Clement, B. M., "Assessing the fidelity of palaeomagnetic records of geomagnetic reversal," *Phil. Trans. R. Soc. Lond.* A**358**, 1049–1064 (2000).

Clement, B. M. and Kent, D. V., "Latitudinal dependency of geomagnetic polarity transition," *Nature* **310**, 488 (1984).

Clement, B. M. and Stixrude, L., "Inner core anisotropy, anomalies in the time-averaged paleomagnetic field, and polarity transition paths," *Earth Planet. Sci. Lett.* **130**, 75–85 (1995).

Clement, B. M., Rodda, P., Smith, E. and Sierra, L., "Recurring transitional geomagnetic field geometries evidence from sediments and lavas," *Geophys. Res. Lett.* **22**, 3171–3174 (1995).

Coe, R. S., Hongre, L. and Glatzmaier, G. A., "An examination of simulated geomagnetic reversals from a palaeomagnetic perspective," *Phil. Trans. R. Soc. Lond.* A**358**, 1141–1170 (2000).

Constable, C. G., "Link between geomagnetic reversal paths and secular variation of the field over the past 5 Myr," *Nature* **358**, 230–232 (1992).

Constable, C. G., "On rates of occurrence of geomagnetic reversals," *Phys. Earth Planet. Inter.* **118**, 181–193 (2000).

Constable, C. G., Tauxe, L. and Parker, R. L., "Analysis of 11 Myr of geomagnetic intensity variation," *J. Geophys. Res.* **103**, 17735–17748 (1998).

Constable, C. G., Johnson, C. L. and Lund, S. P., "Global geomagnetic field models for the past 3000 years: Transient or permanent flux lobes?," *Phil. Trans. R. Soc. Lond.* A**358**, 991–1008 (2000).

Courtillot, V. and Besse, J., "Magnetic field reversals, polar wander, and core–mantle coupling," *Science* **237**, 1140–1147 (1987).

Cox, A., "Lengths of geomagnetic polarity intervals," *J. Geophys. Res.* **73**, 3247 (1968).

Cox, A. and Dalrymple, B., "Statistical analysis of geomagnetic reversal data and the precision of potassium–argon dating," *J. Geophys. Res.* **72**, 2603–2614 (1967).

Cox, A., Doell, R. R. and Dalrymple, G. B., "Reversals of the Earth's magnetic field," *Science* **44**, 1537 (1964).

Creer, K. M. and Ispir, Y., "An interpretation of the behavior of the geomagnetic field during polarity transitions," *Phys. Earth Planet. Inter.* **2**, 283 (1970).

Cronin, M., Tauxe, L., Constable, C., Selkin, P. and Pick, T., "Noise in the quiet zone," *Earth Planet. Sci. Lett.* **190**, 13–30 (2001).

Dormy, E., Valet, J.-P. and Courtillot, V., "Numerical models of the geodynamo and observational constraints," *Geochem. Geophys. Geosyst.* **1**, Paper No. 2000GC000062 (2000).

Egbert, G. D., "Sampling bias in VGP longitudes," *Geophys. Res. Lett.* **19**, 2353–2356 (1992).

Gallet, Y. and Hulot, G., "Stationary and nonstationary behaviour within the geomagnetic polarity time scale," *Geophys. Res. Lett.* **24**, 875–1878 (1997).

Gee, J. S., Cande, S. C., Hildebrand, J. A., Donnelly, K. and Parker, R. L., "Geomagnetic intensity variations over the past 780 kyr obtained from near-seafloor magnetic anomalies," *Nature* **408**, 827–832 (2000).

Gilder, S. and Glen, J., "Magnetic properties of hexagonal closed-packed iron deduced from direct observations in a diamond anvil cell," *Science* **279**, 72–74 (1998).

Glatzmaier, G. A., Coe, R. S., Hongre, L. and Roberts, P. H., "The role of the Earth's mantle in controlling the frequency of geomagnetic reversals," *Nature* **401**, 885–890 (1999).

Gubbins, D., "The distinction between geomagnetic excursions and reversals," *Geophys. J. Int.* **137**, F1–F3 (1999).

Guyodo, Y. and Valet, J. P., "Global changes in geomagnetic intensity during the past 800 thousand years," *Nature* **399**, 249–252 (1999).

Harland, W. B., Armstrong, R. L., Cox, A. V., Craig, L. E., Smith, A. G. and Smith, D. G., *A Geological Time Scale 1989*. Cambridge University Press, Cambridge (1990).

Hengartner, N. W. and Stark, P. B., "Confidence bounds on the probability density of aftershocks," *University of California at Berkeley, Department of Statistics Technical Report, June 1992, #352*, 7pp. (1992a).

Hengartner, N. W. and Stark, P. B., "Conservative finite-sample confidence envelopes for monotone and unimodal densities," *University of California at Berkeley, Department of Statistics Technical Report, September, 1992, #341*, 7pp. (1992b).

Hengartner, N. W. and Stark, P. B., "Finite-sample confidence envelopes for shape-restricted densities," *Ann. Stat.* **23**, 525–550 (1995).

Herrero-Bervera, E., Theyer, F. and Helsley, C. E., "Olduvai onset polarity transition: two detailed paleomagnetic records from North Central Pacific deep-sea sediments," *Nature* **322**, 159–162 (1987).

Hilgen, F. J., "Astronomical calibration of Gauss to Matuyama sapropels in the Mediterranean and implications for the geomagnetic polarity time scale," *Earth Planet. Sci. Lett.* **104**, 226–244 (1991).

Hillhouse, J. and Cox, A., "Brunhes–Matuyama polarity transition," *Earth Planet. Sci. Lett.* **29**, 51–64 (1976).

Hoffman, K. A., "Temporal aspects of the last reversal of Earth's magnetic field," *Phil. Trans. R. Soc. Lond.* A**358**, 1181–1190 (2000).

Hollerbach, R. and Jones, C., "A geodynamo model incorporating a finitely conducting inner core," *Phys. Earth Planet. Inter.* **75**, 317–327 (1993a).

Hollerbach, R. and Jones, C., "Influence of Earth's inner core on geomagnetic fluctutations and reversals," *Nature* **365**, 541–543 (1993b).

Hollerbach, R. and Jones, C., "On the magnetically stabilising role of the Earth's inner core," *Phys. Earth Planet. Inter.* **87**, 171–181 (1995).

Jacobs, J. A., *Reversals of the Earth's Magnetic Field*. Cambridge University Press, Cambridge (1994).

Johnson, R. G., "Brunhes–Matuyama magnetic reversal dated at 790,000 yr BP by marine astronomical correlations," *Quatern. Res.* **17**, 135 (1982).

Johnson, H. P., Van Patten, D., Tivey, M. and Sager, W., "Geomagnetic polarity reversal rate for the Phanerozoic," *Geophys. Res. Lett.* **22**, 231–234 (1995).

Katari, K. and Bloxham, J., "Effects of sediment aggregate size on DRM intensity: a new theory," *Earth Planet. Sci. Lett.* **186**, 113–122 (2001).

Katari, K. and Tauxe, L., "Effects of pH and salinity on the intensity of magnetization in redeposited sediments," *Earth Planet. Sci. Lett.* **181**, 489–496 (2000).

Katari, K., Tauxe, L. and King, J., "A reassessment of post-depositional remanent magnetism: preliminary experiments with natural sediments," *Earth Planet. Sci. Lett.* **183**, 147–160 (2000).

Kent, D. V. and Carlut, J. C., "Paleointensity record in zero-age submarine basaltic glasses: testing a new dating technique for recent MORBs," *Earth Planet. Sci. Lett.* **183**, 389–401 (2000).

Kent, D. V. and Smethurst, M. A., "Shallow bias of paleomagnetic inclinations in the Paleozoic and Precambrian," *Earth Planet. Sci. Lett.* **160**, 392–402 (1998).

Laj, C., Mazaud, A., Weeks, R., Fuller, M. and Herrero-Bervera, E., "Geomagnetic reversal paths," *Nature* **351**, 447 (1991).

Langereis, C. G., van Hoof, A. A. M. and Rochette, P., "Longitudinal confinement of geomagnetic reversal paths as a possible sedimentary artefact," *Nature* **358**, 226–229 (1992).

Langereis, C. G., Dekkers, M. J., de Lange, G. J., Paterne, M. and van Santvoort, P. J. M., "Magneto-stratigraphy and astronomical calibration of the last 1.1 Myr from an eastern Mediterranean piston core and dating of short events in the Brunhes," *Geophys. J. Int.* **129**, 75–94 (1997).

Larsen, R. and Olson, P., "Mantle plumes control magnetic reversal frequency," *Earth Planet. Sci. Lett.* **107**, 437–447 (1991).

Levi, S., Audunsson, H., Duncan, R. A., Kristjansson, L., Gillot, P. Y. and Jakobsson, S. P., "Late Pleistocene geomagnetic excursion in Icelandic lavas – confirmation of the Laschamp excursion," *Earth Planet. Sci. Lett.* **96**, 443–457 (1990).

Lock, J. and McElhinny, M. W., "Special issue – the global paleomagnetic database design, installation, and use with Oracle," *Surv. Geophys.* **12**, 5–91 (1991).

Loper, D., "On the correlation between mantle plume flux and the frequency of reversals of the geomagnetic field," *Geophys. Res. Lett.* **19**, 25–28 (1992).

Loper, D. and McCartney, K., "Mantle plumes and the periodicity of magnetic field reversals," *Geophys. Res. Lett.* **13**, 1525–1528 (1986).

Love, J. J., "Paleomagnetic volcanic data and geometric regularity of reversals and excursions," *J. Geophys. Res.* **103**, 12435–12452 (1998).

Love, J. J. and Mazaud, A., "A database for the Matuyama–Brunhes magnetic reversal," *Phys. Earth Planet. Inter.* **103**, 207–245 (1997).

Lund, S. P. *et al.*, "Geomagnetic field excursions occurred often during the last million years," *EOS, Trans. AGU* **79**, 178–179 (1998).

McElhinny, M. W., "Geomagnetic reversals during the Phanerozoic," *Science* **172**, 157–159 (1971).

McElhinny, M. W. and Lock, J., "Global paleomagnetic database suppplement number one – update to 1992," *Surv. Geophys.* **14**, 303–329 (1993).

McFadden, P. L. and Merrill, R. T., "Lower mantle convection and geomagnetism," *J. Geophys. Res.* **89**, 3354–3362 (1984).

McFadden, P. L. and Merrill, R. T., "Asymmetry in the reversal rate before and after the Cretaceous Normal Polarity Superchron," *Earth Planet. Sci. Lett.* **149**, 43–47 (1997).

McFadden, P. L. and Merrill, R. T., "Evolution of the geomagnetic reversal rate since 160 Ma: Is the process continuous?," *J. Geophys. Res.* **105**, 28455–28460 (2000).

McFadden, P. L., Merrill, R. T., McElhinny, M. W. and Lee, S., "Reversals of the Earth's magnetic field and temporal variations of the dynamo families," *J. Geophys. Res.* **96**, 3923–3933 (1991).

McMillan, D. G., Constable, C. G., Parker, R. L. and Glatzmaier, G. A., "A statistical analysis of magnetic field from some dynamo simulations," *Geophys. Geochem. Geosyst.* **2**, 10.129/2000GC000130 (2001).

Mazaud, A., "An attempt at reconstructing the geomagnetic field at the core–mantle boundary during the Upper Olduvai polarity transition (1.6 Myear)," *Phys. Earth Planet. Inter.* **90**, 211–219 (1995).

Merrill, R. T. and McFadden, P. L., "Geomagnetic polarity transitions," *Rev. Geophys.* **37**, 201–226 (1999).

Merrill, R. T., McElhinny, M. W. and McFadden, P. L. *The Magnetic Field of the Earth: Paleomagnetism, the Core and The Deep Mantle*. Academic Press, San Diego, CA (1996).

Nowaczyk, N. R. and Antonow, M., "High-resolution magnetostratigraphy of four sediment cores from the Greenland Sea .1. Identification of the Mono Lake excursion, Laschamp and Biwa I Jamaica geomagnetic polarity events," *Geophys. J. Int.* **131**, 310–324 (1997).

Olsen, N. *et al.* (25 authors), "Ørsted initial field model," *Geophys. Res. Lett.* **27**, 3607–3610 (2000).

Pavlov, V. and Gallet, Y., "Middle Cambrian high magnetic reversal frequency (Kulumbe River section, northwestern Siberia) and reversal behaviour during the Early Palaeozoic," *Earth Planet. Sci. Lett.* **185**, 173–183 (2001).

Pouliquen, G., Gallet, Y., Patriat, P., Dyment, J. and Tamura, C., "A geomagnetic record over the last 3.5 million years from deep-tow magnetic anomaly profiles across the Central Indian Ridge," *J. Geophys. Res.* **106**, 10941–10960 (2001).

Prévot, M. and Camps, P., "Absence of preferred longitude sectors for poles from volcanic records of geomagnetic reversals," *Nature* **366**, 53–56 (1993).

Prévot, M., Derder, M. E., McWilliams, M. and Thompson, J., "Intensity of the Earth's magnetic field: evidence for a Mesozoic dipole low," *Earth Planet. Sci. Lett.* **97**, 129–139 (1990).

Selkin, P. and Tauxe, L., "Long term variations in geomagnetic field intensity," *Phil. Trans. R. Soc. Lond.* **A358**, 869–1223 (2000).

Shackleton, N. J., Berger, A. and Peltier, W. R., "An alternative astronomical calibration of the lower Pleistocene timescale based on ODP site 677," *Trans. Roy. Soc. Edinburgh, Earth Sci.* **81**, 251–261 (1990).

Shao, J. C., Fuller, M., Tanimoto, T., Dunn, J. R. and others, "Spherical harmonic analyses of paleo-magnetic data: the time-averaged geomagnetic field for the past 5 Myr and the Brunhes–Matuyama reversal," *J. Geophys. Res.* **104**, 5015–5030 (1999).

Singer, B. A. and Pringle, M. S., "Age and duration of the Matuyama–Brunhes geomagnetic polarity reversal from ^{40}Ar/^{39}Ar incremental heating analyses of lavas," *Earth Planet. Sci. Lett.* **139**, 47–61 (1996).

Singer, B. A., Hoffman, K. A., Chauvin, A., Coe, R. S. and Pringle, M. S., "Dating transitionally magnetized lavas of the late Matuyama Chron: toward a new ^{40}Ar/^{39}Ar timescale of reversals and events," *J. Geophys. Res.* **104**, 679–693 (1999).

Tarduno, J. A., Cottrell, R. D. and Smirnov, A. V., "High geomagnetic intensity during the mid-Cretaceous from Thellier analyses of single plagioclase crystals," *Science* **291**, 1779–1783 (2001).

Tauxe, L., "How non-dipolar was the ancient geomagnetic field?," *EOS, Trans. AGU* **80**(46), F304 (1999).

Tauxe, L., Deino, A. D., Behrensmeyer, A. K. and Potts, R., "Pinning down the Brunhes/Matuyama and upper Jaramillo boundaries: a reconciliation of orbital and isotopic time scales," *Earth Planet. Sci. Lett.* **109**, 561–572 (1992).

Tauxe, L., Herbert, T., Shackleton, N. J. and Kok, Y. S., "Astronomical calibration of the Matuyama–Brunhes boundary: consequences for magnetic remanence acquisition in marine carbonates and the Asian loess sequences," *Earth Planet. Sci. Lett.* **140**, 133–146 (1996).

Thouveny, N. and Creer, K. M., "On the brevity of the Laschamp excursion," *Bull. Soc. Géol. France* **163**, 771–780 (1992).

Valet, J.-P. and Laj, C., "Invariant and changing transitional field configurations in a sequence of geomagnetic reversals," *Nature* **311**, 352 (1984).

Valet, J.-P. and Meynadier, L., "Geomagnetic field intensity and reversals during the last four million years," *Nature* **366**, 234–238 (1993).

Valet, J.-P., Tauxe, L. and Clement, B. M., "Equatorial and mid-latitude records of the last geomagentic reversal from the Atlantic Ocean," *Earth Planet. Sci. Lett.* **94**, 371–384 (1989).

Valet, J. P., Tucholka, P., Courtillot, V. and Meynadier, L., "Palaeomagnetic constraints on the geometry of the geomagnetic field during reversals," *Nature* **356**, 400–407 (1992).

Verosub, K., "Geomagnetic excursions and their paleomagnetic record," *Rev. Geophys. Space Phys.* **15**, 145–155 (1977).

Vine, F. and Matthews, D. H., "Magnetic anomalies over oceanic ridges," *Nature* **199**, 947–949 (1963).

Vlag, P., Thouveny, N., Williamson, D., Rochette, P. and Benatig, F., "Evidence for a geomagnetic field excursion recorded in the sediments of Lac St Front, France – a link with the Laschamp excursion," *J. Geophys. Res.* **101**, 28211–28230 (1996).

Worm, H.-U., "A link between geomagnetic reversals and events and glaciations," *Earth Planet. Sci. Lett.* **147**, 55–67 (1997).

5 Energy fluxes and ohmic dissipation in the earth's core[*]

Paul H. Roberts[1], *Christopher A. Jones*[2] *and*
Arthur R. Calderwood[3]

[1] *Institute of Geophysics and Planetary Physics, University of California, Los Angeles,
CA 90095, USA*
[2] *School of Mathematical Sciences, University of Exeter, Exeter, EX4 4QE, UK*
[3] *Physics Department, University of Nevada-Las Vegas, Las Vegas, NV 89154, USA*

Simulations have shown that convection-driven geodynamo models can produce reversing magnetic fields of a similar strength to those of the geomagnetic field. These simulations suggest that the ohmic dissipation arising from the geodynamo is not due solely to the largest scale harmonics of the magnetic field, but is dominated by contributions from smaller scale components. It is argued that, in order to produce a vigorous reversing dynamo resembling the geodynamo, ohmic dissipation of between 1 and 2 TW is required.

The energy and entropy fluxes through the core are investigated to see under what circumstances the required ohmic dissipation can be produced. The estimates for these fluxes given in Braginsky and Roberts (1995) are revised in the light of recent theoretical and experimental advances in high pressure physics. In the absence of radioactive heating in the core, the available thermal and gravitational energy sources can produce more than 1 TW of ohmic dissipation provided the inner core is not significantly older than 1.2 Gyr. If the inner core is much older than this, the thermal and gravitational energy sources are released at too slow a rate to give the required amount of ohmic dissipation. Paleomagnetic evidence suggests that a geomagnetic field of approximately today's strength has been in existence for over 3 Gyr, perhaps before the inner core came into existence. Estimates of the available energy sources based on current estimates of the heat flux passing through the core–mantle boundary suggest that they may be insufficient to sustain such a field. We consider the possibility that significant radioactivity in the core could resolve this difficulty.

1. Introduction

Many bodies in the solar system are known to have dense central cores, examples being Earth, Venus, Mars, Mercury, Ganymede. Some of these bodies (e.g. Earth, Mercury, Ganymede) create their own magnetic fields by dynamo action in their cores; others (e.g. Venus, Mars) do not. In the case of the Earth, the principal example used in this article, there are strong indications that the core is composed of an iron-rich alloy; the cores of the other terrestrial planets and satellites are usually taken to have a similar metallic composition. The salient property of such materials (when in a melted state) is a magnetic diffusivity η_M that is large

---

[*] This manuscript was accepted by the Editors on May 31, 2001.

compared with all other diffusivities, such as the kinematic viscosity ν_M and the thermal diffusivity κ_M, where the suffix M, standing for "molecular," is added to avoid confusion with the "turbulent" diffusivities to be introduced later, for which T is used. The Prandtl numbers,

$$P_m \equiv \nu_M/\eta_M, \quad q \equiv \kappa_M/\eta_M, \tag{1.1}$$

are therefore very small: values quoted for the Earth's core are typically $P_m \sim 10^{-6}$ and $q \sim 10^{-5}$.

To maintain a magnetic field successfully, the core flow speed, of order $\mathcal{V}$ (say), must be sufficently large, as measured by the magnetic Reynolds number,

$$Rm \equiv \mathcal{V}\mathcal{L}/\eta_M, \tag{1.2}$$

where $\mathcal{L}$ is the depth of the fluid core. A characteristic value of Rm for a working dynamo is taken here to be 100, which by Eq. (1.1) implies that

$$Re \equiv \mathcal{V}\mathcal{L}/\nu_M, \quad Pe \equiv \mathcal{V}\mathcal{L}/\kappa_M, \quad Mp \equiv \mathcal{V}\mathcal{L}/D_M, \tag{1.3}$$

are all very large. We have here included, with the (kinetic) Reynolds number $Re \sim 10^8$ and the Péclet number $Pe \sim 10^7$, the mass Péclet number $Mp \sim 10^{11}$ which is a dimensionless measure of the diffusivity, D, between the different chemical constituents present in the core. The predominant composition of the Earth's core is unknown, FeS, FeSi and FeO having all been advocated. Recent work (Price *et al.* presented as a poster in this meeting) indicates that a trinary alloy is more realistic. To generalize the following from two to three (or indeed many) components is easy, but adds complications without compensating enlightenment; for simplicity, we shall model the fluid as a binary alloy. The large values of Re, Pe and Mp indicate that molecular diffusion of large-scale momentum, heat and composition are totally negligible, and that these quantities must be transported predominantly by small-scale eddies which, while they may contain very little kinetic energy, are far more effective diffusers than molecules.

It is impossible now or in the foreseeable future to resolve numerically the small-scale eddies that diffuse the large-scale fields. For instance, in the case of Earth, it would be necessary to resolve length scales of the order of $\ell \sim 1$ km, in a body of fluid of depth $\mathcal{L} \sim 2,900$ km, and to follow time-scales of order $\tau \sim 300$ s in a system where the overturning time of large eddies is of order 300 yrs ($= 10^{10}$ s). Lacking anything better, we employ the popular two-scale approach in which the flow **V** is modeled as a superposition of a small-scale turbulent velocity **V**$'$ and a large-scale laminar motion $\overline{\textbf{V}}$, so that $\textbf{V} = \overline{\textbf{V}} + \textbf{V}'$. We then follow Osborne Reynolds who, likening the nearly random motion of eddies to the nearly random motion of molecules, argued that the effect of turbulence on the transport of large-scale momentum could be well represented by replacing the molecular kinematic viscosity ν_M by a much greater "turbulent viscosity," ν_T. According to the Reynolds analogy, the flux, at position **x** and time t, of large-scale momentum created by the small-scale eddies depends only on the rates of strain of the large-scale velocity, $\overline{\textbf{V}}(\textbf{x}, t)$, at the same **x** and t. In other words, the analogy leads to a *local* description of turbulence, in which the inertial term $\overline{\textbf{V}' \cdot \nabla \textbf{V}'}$ in the equation governing $\overline{\textbf{V}}$ is represented by $-\nu_T \nabla^2 \overline{\textbf{V}}$. This has the effect of replacing ν_M by $\nu_{total} = \nu_T + \nu_M$ in the equation governing $\overline{\textbf{V}}$, but since $\nu_T \gg \nu_M$ it is acceptable to set $\nu_{total} = \nu_T$.

The hugeness of Pe and Mp suggests that, in the case of the specific entropy S and the composition ξ (i.e. the mass fraction of the light component in the alloy), the convective fluxes $\mathbf{I}^S = \rho_a \overline{S'\mathbf{V}'}$ and $\mathbf{I}^\xi = \rho_a \overline{\xi'\mathbf{V}'}$ of large-scale entropy and composition created by the small-scale motions may be replaced respectively by $-\rho_a \kappa_T \nabla \overline{S}$ and $-\rho_a D_T \nabla \overline{\xi}$ where, again, the diffusivities κ_T and D_T so greatly exceed κ_M and D_M that they can be used to replace the more precise κ_{total} and D_{total}. Moreover, Braginsky and Roberts (1995) argue that, since S and ξ are extensive variables that are transported by turbulent eddies in the same way, $D_T = \kappa_T$.

The Reynolds analogy implicity assumes that the eddy motions, like the molecular motions, are approximately isotropic. It has been cogently argued however by Braginsky (1964) and Braginsky and Meytlis (1990) that core turbulence is far from isotropic; see also Braginsky and Roberts (1995) and St Pierre (1996). Thus, ν_T and κ_T should be replaced by tensors. This has never been done in numerical simulations, although a different (and undesirable) anisotropy is present in many geodynamo simulations with the introduction of hyperviscosity, which enhances diffusion in the horizontal directions but not in the radial direction. This is done for purely pragmatic reasons; the anisotropy of core turbulence is almost certainly determined by the directions of the prevailing mean field, $\overline{\mathbf{B}}$, and the rotation of the system, $\mathbf{\Omega}$, and not by the local direction of gravity, $\mathbf{g}$. In what follows we shall follow current practice by ignoring diffusive anisotropies. We shall sometimes use the terms "macroscale" and "microscale" instead of "large-scale" and "small-scale."

A perplexing question concerns the effect of turbulence on the mean magnetic field $\overline{\mathbf{B}}$, "Does microscale induction significantly affect the macroscales?" Put another way, "Must the full paraphenalia of mean field electrodynamics, including a turbulent magnetic diffusivity η_T and perhaps an α-effect, be invoked?"

On the one hand, the answer is clearly "No." Although the macroscale magnetic Reynolds number Rm is necessarily large, it is apparently not very large. If $Rm = 100$ and $\ell/\mathcal{L} = 1/3,000$ (see above) and, if the characteristic turbulent velocity $\mathcal{V}'$ is of the same order as the characteristic macroscale velocity $\mathcal{V}$, the microscale Reynolds number, $Rm' \equiv \mathcal{V}'\ell/\nu_M$, is only about 0.03. Moreover, a drastic increase in η from the currently favored value of $\eta_M \approx 2\,\text{m}^2\,\text{s}^{-1}$ to $\eta_{\text{total}} = \eta_T + \eta_M$ where $\eta_{\text{total}} \gg \eta_M$ would result in a corresponding reduction in the time scale of the macroscale field for which, in the case of the Earth, there is no observational evidence. The study of paleomagnetism has shown that a polarity reversal of the Earth's field is accomplished in a time of the order of 5,000 yr or greater, which is comparable with the free-decay time, $\tau_\eta \equiv \mathcal{L}^2/\pi^2 \eta_M \approx 20,000\,\text{yr}$, of the dipole field in the absence of turbulence. This situation may be contrasted with that for the Sun, for which η_T exceeds η_M by a very large factor, and for which the time scale of magnetic activity is enormously less than τ_η.

On the other hand, all numerical simulations that successfully imitate the Earth by exhibiting polarity reversals have been driven strongly enough to be irregular; see for example Glatzmaier and Roberts (1995a,b, 1996a,b) and Sarson and Jones (1999). The fields fluctuate in time, and there is a suspicion that a polarity reversal is merely another fluctuation that happens to be rather larger than the others. Such irregularities in $\overline{\mathbf{B}}$ suggest that fluctuations in $\mathbf{B}'$ may be significant also, and especially so because the associated electric current density $\mathbf{J}' = \nabla \times \mathbf{B}'/\mu_0 \sim \mathcal{B}'/\mu_0 \ell$ (where μ_0 is the magnetic permeability) is enhanced relative to $\overline{\mathbf{J}} = \nabla \times \overline{\mathbf{B}}/\mu_0 \sim \overline{\mathcal{B}}/\mu_0 \mathcal{L}$ by the factor $\mathcal{L}/\ell$, so that $\mathcal{J}'/\overline{\mathcal{J}} \sim (\mathcal{L}/\ell)\mathcal{B}'/\overline{\mathcal{B}} \sim 3000\mathcal{B}'/\overline{\mathcal{B}}$. This means that the Joule losses per unit volume, $q^J = \mu_0 \eta_M \mathbf{J}^2$, of the electric currents $\mathbf{J}$ may have a significant microscale part $\overline{q^{J'}} = \mu_0 \eta_M \overline{(\mathbf{J}')^2}$ in addition to the obvious macroscale contribution $\overline{q}^J = \mu_0 \eta_M \overline{\mathbf{J}}^2$. This is confirmed in Section 2, where we show that the smaller scale components of the magnetic field make a bigger contribution to the Joule losses than the lowest order components such as the dipole field. We are able to glean information on how rapidly the field components fall off with increasing wavenumber, and hence to estimate the total Joule dissipation, albeit rather approximately. Our arguments suggest that, although the turbulent magnetic diffusivity exceeds the molecular magnetic diffusivity, it does not do so by a large factor.

Our estimate of q^J has a strong bearing on the thermodynamic arguments to be given in Sections 4 and 5. These are based on a model of the core which is developed in Section 2. In Section 6 we present a case, based purely on considerations of core thermodynamics for the existence of significant radioactivity in the core.

2. Ohmic dissipation

2.1. Order of magnitude estimates

Glatzmaier and Roberts (1996a) estimated that the contribution, $\mathcal{Q}^{\overline{\mathcal{J}}}$, to the total ohmic dissipation $\mathcal{Q}^J$ made by the large scale fields in their simulation was on average about

$$\mathcal{Q}^{\overline{\mathcal{J}}} = 0.3 \, \text{TW}. \tag{2.1}$$

A crude argument suggests that this is much less than $\mathcal{Q}^{J'}$. Fluctuations of magnetic field $\mathbf{B}'$ on a short length-scale ℓ produced by velocity fluctuations $\mathbf{V}'$ acting on a large scale mean field $\overline{\mathbf{B}}$ satisfy

$$\nabla \times (\mathbf{V}' \times \overline{\mathbf{B}}) \sim \eta_M \nabla^2 \mathbf{B}', \quad \text{or} \quad B' \sim \frac{V'\ell}{\eta_M} \overline{B}. \tag{2.2, 2.3}$$

The large- and small-scale dissipations are then

$$\mathcal{Q}^{\overline{\mathcal{J}}} \sim \frac{\eta_M \overline{B}^2}{\mu_0 \mathcal{L}^2} \mathcal{V}_{\text{FOC}}, \quad \mathcal{Q}^{J'} \sim \frac{\eta_M B'^2}{\mu_0 \ell^2} \mathcal{V}_{\text{FOC}}, \tag{2.4}$$

where $\mathcal{V}_{\text{FOC}}$ is the volume of the fluid outer core (FOC). If we further assume that the fluctuating velocities $\mathbf{V}'$ are of the same order as the large-scale velocity $\mathcal{V}$, we find that

$$\frac{\mathcal{Q}^{J'}}{\mathcal{Q}^{\overline{\mathcal{J}}}} \sim \frac{B'^2 \mathcal{L}^2}{\overline{B}^2 \ell^2} \sim \left(\frac{\mathcal{V}\mathcal{L}}{\eta_M}\right)^2 = Rm^2. \tag{2.5}$$

Since Rm is at least 100 in the core, this suggests $\mathcal{Q}^{J'} \sim 10^4 \mathcal{Q}^{\overline{\mathcal{J}}}$. Equation (2.1) would then imply that $\mathcal{Q}^J \approx \mathcal{Q}^{J'} \sim 3{,}000 \, \text{TW}$, which is clearly far too large.

Some reduction in $\mathcal{Q}^{J'}$ could be obtained if turbulence fills only part of the FOC, or is strongly intermittent. However, the estimate (2.3) can also be questioned. It is essentially kinematic in origin; Lorentz forces may reduce the cross-field velocity, hence reducing $\mathbf{V}' \times \overline{\mathbf{B}}$. Some evidence for this is present in plane-layer dynamo calculations (Jones and Roberts, 2000). One may also object that Eq. (2.1) comes from a simulation that includes "intermediate" length scales (down to around 100 km), and so it already includes some of the smaller scale dissipation. In other words, a substantial part of $\mathcal{Q}^{\overline{\mathcal{J}}}$ should be re-assigned to $\mathcal{Q}^{J'}$.

There is a common perception that the geodynamo is of strong field type, meaning that the Coriolis and Lorentz forces associated with the large-scale fields are of the same order of magnitude, so that $\overline{J}$ is of order $2\Omega\rho\mathcal{V}/\overline{B}$. The simulations that led to Eq. (2.1) gave typical field strengths of order $10^{-2} \, \text{T}$, and, taking $\rho = 10^4 \, \text{kg m}^{-3}$ and $\mathcal{V} = 10^{-4} \, \text{m s}^{-1}$, we find that $\overline{J}$ is of order $1.4 \times 10^{-2}\,\mathrm{A\,m^{-2}}$. Taking $\eta_M = 2\,\mathrm{m^2\,s^{-1}}$, we obtain $\mathcal{Q}^{\overline{J}} = 0.08\,\mathrm{TW}$. This suggests that 75% of the 0.3 TW given in Eq. (2.1) belongs in $\mathcal{Q}^{J'}$. Glatzmaier and Roberts (1996a) obtained Eq. (2.1) from the scales that they resolved in their computation, which included all spherical harmonics up to degree 21, corresponding to length scales of about 100 km. They described the remainder as "turbulent dissipation" and assigned it to their unresolved scale; they assessed it at about 0.1 TW. Their simulation, in common with most of the other simulations of strongly driven dynamos, employed hyperdiffusion, which artificially damps out the shorter length scales, and decreases $\mathcal{Q}^{J'}$ artificially relative to $\mathcal{Q}^{\overline{J}}$. This substantially increases the rate at which the magnetic energy spectrum decreases with increasing harmonic number. (Compare the spectrum from their simulation with that of the highly resolved model reported in Roberts and Glatzmaier (2000a), which does not employ hyperdiffusion.) It is therefore likely that the contribution made to $\mathcal{Q}^{J}$ by their unresolved scales was unrepresentatively small and that, if hyperdiffusion had been eliminated, $\mathcal{Q}^{J}$ would have been much larger than 0.4 TW.

Although scale separation is convenient and often used, it is an oversimplification. In reality "intermediate scales" are significant, as we shall demonstrate in Section 2.2. This leads to an ambiguity in the interpretation of results like Eq. (2.1) in two-scale language.

2.2. Large-scale dissipation in the core

We first consider the *minimum* dissipation in the core associated with the observed components of the geomagnetic field. Only the poloidal part of the field is visible at the surface, so no toroidal field is included in this minimum dissipation calculation. Gubbins (1977) showed that the minimum dissipation associated with the geomagnetic field is

$$Q^J = \sum_{n=1}^{\infty} q_n, \quad q_n = \frac{\eta_M r_{CMB}}{\mu_0} \frac{(2n+1)(2n+3)}{n} R_n, \tag{2.6}$$

where q_n is the dissipation from the spherical harmonics of order n, and

$$R_n = \left(\frac{r_E}{r_{CMB}}\right)^{2n+4} (n+1) \sum_{m=0}^{n} \left[\left(g_n^m\right)^2 + \left(h_n^m\right)^2\right] \tag{2.7}$$

is the Mauersberger–Lowes spectrum extrapolated to the core surface (see, e.g. Langel, 1987), r_E and r_{CMB} being the radii of the Earth and of the core–mantle boundary (CMB), respectively; g_n^m and h_n^m are the usual Gauss coefficients that multiply Schmidt-normalized associated Legendre functions in the expansion of the magnetic potential at the Earth's surface (see e.g. Backus *et al.*, 1996). Using MAGSAT values of the Gauss coefficients and our standard estimate of $\eta = 2\,\mathrm{m^2\,s^{-1}}$, we find that

$$q_1 = 5.82\,\mathrm{MW} \quad \text{and} \quad q_2 = 0.75\,\mathrm{MW}. \tag{2.8}$$

However, as n increases past 2, the dissipation starts to increase with n, reaching a maximum at q_8 of 1.5 MW. This strongly suggests that the "intermediate" scales contribute significantly to the total dissipation. Observations indicate that the Mauersberger–Lowes spectrum R_n at the CMB is well approximated for $n \geq 3$ by

$$R_n = 1.51 \times 10^{-8} \exp(-0.1n)\,\mathrm{T^2} \tag{2.9}$$

and using Eqs (2.6)–(2.8) we can derive an estimate of the minimum total dissipation associated with the observed field:

$$Q^J_{\text{min}} = \sum_{n=1}^{\infty} q_n = q_1 + q_2 + \frac{\eta_M r_{\text{CMB}}}{\mu_0} \sum_{n=3}^{\infty} \frac{(2n+1)(2n+3)}{n} R_n, \tag{2.10}$$

which by Eq. (2.9) gives $Q^J_{\text{min}} \approx 44\,\text{MW}$. This agrees well with the estimate of Roberts and Glatzmaier (2000b): $Q^J_{\text{min}} \approx 43\,\text{MW}$. It should be noted that a significant part of Q^J_{min} (about 25 MW) is made by modes with $n > 12$, which is beyond the testable range of Eq. (2.9), because the core field signal is then swamped by crustal field contributions.

The actual magnetic field has to satisfy the MHD equations, a constraint not imposed in the derivation of Eq. (2.6). A somewhat more realistic estimate can be obtained from the dissipation associated with poloidal and toroidal decay modes (see, e.g. Moffatt, 1978). The ohmic dissipation associated with the poloidal modes is bounded below by the dissipation of the decay modes that have the same power spectra at the CMB. From this consideration, we see that q_n cannot be smaller than

$$q_n = \frac{2\pi \eta_M r_{\text{CMB}}}{\mu_0} \frac{\alpha_{n-1}^4}{n(2n+1)} R_n, \tag{2.11}$$

where α_n is the smallest positive zero of the spherical Bessel function, $j_n(\alpha)$. This gives, as lower limits,

$$q_1 = 79.5\,\text{MW} \quad \text{and} \quad q_2 = 11.0\,\text{MW}. \tag{2.12}$$

We can again use the Mauersberger–Lowes spectrum (2.9) to estimate the total dissipation. As with Eq. (2.6), the dissipation increases with n when $n \geq 3$, reaching a maximum of 36.2 MW, this time of $n = 14$. We can use the first three terms of the asymptotic formula for the first zero of the Bessel function (e.g. Abramowitz and Stegun, 1964):

$$\alpha_n \sim (n + \tfrac{1}{2}) + 1.85576(n + \tfrac{1}{2})^{1/3} + 1.03315(n + \tfrac{1}{2})^{-1/3} \tag{2.13}$$

to obtain an alternative, and more stringent, lower bound

$$Q^J_{\text{min}} = \sum_{n=1}^{\infty} q_n \approx q_1 + q_2 + \sum_{n=3}^{\infty} q_n \approx 1.32\,\text{GW}, \tag{2.14}$$

where now Eqs (2.11)–(2.13) are used to evaluate Eq. (2.14). Note that the higher harmonics make an even greater contribution to the dissipation than was the case for the "minimizing" field. This is because, as Eq. (2.13) shows, α_n is linear with n in the large n limit, so the dissipation is asymptotically proportional to $n^2 \exp(-0.1n)$ for the decay modes, but only proportional to $n \exp(-0.1n)$ for the "minimizing" field. Of course, the caveats mentioned above about the uncertainty of the Mauersberger–Lowes spectrum at $n = 14$ apply with even more force in this case. Nevertheless, it is interesting that this spectrum suggests that the higher n modes dissipate over 16 times the energy than the $n = 1$ (dipole) mode.

In view of the fact that the observed spectrum (2.9) is based only on observations of modes with $n < 12$, it is of interest to compare Eq. (2.9) with spectra derived from geodynamo simulations. A recent simulation (Roberts and Glatzmaier, 2000a), using a very large number of modes but run for a comparatively short time, gave a spectrum of the form (2.9) but with an exponent -0.055 instead of -0.1. Although the run was short (only about 1,500 yr), this is sufficient to establish the power spectrum, and harmonics up to $n = 239$ were included. The value of viscosity used naturally had to exceed that of the Earth by a considerable factor, but nevertheless it is small enough for the great bulk of the dissipation to be ohmic rather than viscous. We therefore consider how Eq. (2.14) is affected if the spectrum,

$$R_n = 1.51 \times 10^{-8} \exp(-0.055n)\, T^2, \tag{2.15}$$

is used in place of Eq. (2.9). We have left the pre-factor unchanged, as this does not greatly affect the lower-order modes, which are the most confidently known. The effect is to increase the dissipation in Eq. (2.14) by over a factor 4 to 5.77 GW, with the peak now occurring at $n = 29$ rather than $n = 14$.

We can also derive some estimates for the losses from toroidal modes, though here we have to rely on numerical simulations to estimate the strength of the toroidal fields. The dissipation can be computed directly from these simulations, but as this is not always done it is of interest to relate the dissipation directly to the average field strengths in the simulations, which are usually given.

The toroidal decay modes can be written

$$\mathbf{B} = \nabla \times [T_{nm}\, j_n(kr)\, P_n^m(\cos\theta) \exp(im\phi)\mathbf{r}], \tag{2.16}$$

where $k = \alpha/r_{\mathrm{CMB}}$. The root mean square magnetic field associated with this mode is

$$\overline{B}^2 = \frac{1}{V_{\mathrm{core}}} \int_{\mathrm{core}} \mathbf{B}^2\, d^3x = \frac{3n(n+1)}{2(2n+1)} \left[j_n'(\alpha)\right]^2 |T_{nm}|^2, \tag{2.17}$$

where V_{core} is the volume of the entire core. The right-hand side of Eq. (2.17) is proportional to the ohmic dissipation produced by the mode, which is

$$q_n^m = \frac{\eta_M r_{\mathrm{CMB}}}{\mu_0} \frac{2n(n+1)}{2n+1} \left[\alpha j_n'(\alpha)\right]^2 |T_{nm}|^2 = \frac{4\pi r_{\mathrm{CMB}} \eta_M \alpha^2}{3\mu_0} \overline{B}^2. \tag{2.18}$$

The lowest-order toroidal mode with dipole symmetry (B_ϕ antisymmetric about the equator) is $n = 2$, $m = 0$ for which the smallest nontrivial α is 5.7635. With our standard estimates this gives $q_2 = 7.7 \times 10^{14}\overline{B}^2$ W. The Glatzmaier–Roberts simulations had a toroidal field of order 100 G, which is 0.01 T, giving a dissipation $q_2 \sim 0.077$ TW, about one-quarter of the dissipation (2.1). If the spectrum for the toroidal modes follows the same power law (2.15) as the poloidal modes (with a different prefactor) the total power consumption will be almost twenty times greater, that is, about 1.5 TW.

The dissipation coming from the toroidal modes seems to be much larger than that from the poloidal modes (2.14). However, this is because our estimate for the poloidal field comes from the observed Gauss coefficients, which correspond to a field strength of only 0.05–0.1 mT. All simulations suggest that the poloidal field is considerably stronger than this, with most of the poloidal field forming closed field lines that do not escape from the FOC, and hence do not contribute to the Gauss coefficients. We estimate that the toroidal and poloidal field together give a total ohmic power loss in the range,

$$1\,\mathrm{TW} < Q^J < 2\,\mathrm{TW}. \tag{2.19}$$

In most of what follows, Eq. (2.19) is more significant than Eq. (2.1); the latter is of interest only in Section 5.

3. The adiabatic state

3.1. Seismically determined variables

In this section, we describe how the estimates to be presented in Section 4 were obtained. We suppose that the fluid alloy is, to the first approximation, in a well-mixed hydrostatic state under its own self-produced gravitational field $\mathbf{g}_a$,

$$\nabla P_a = \rho_a \mathbf{g}_a, \quad \nabla S_a = \mathbf{0}, \quad \nabla \xi_a = \mathbf{0}, \tag{3.1, 3.2, 3.3}$$

where P is pressure and ρ is density; the suffix "a" refers to the fact that this is an adiabatic state, by Eq. (3.2). Other (intensive) variables are not homogenized by the mixing. In the case of the temperature T, for example,

$$T_a^{-1} \nabla T_a = \gamma \mathbf{g}_a / u_P^2, \quad \text{where } \gamma = \alpha u_P^2 / c_p \tag{3.4}$$

is the Grüneisen parameter, u_P is the speed of sound, α is the coefficient of thermal expansion, and c_p is the specific heat at constant pressure.

It is possible to use Eqs (3.1)–(3.3) to construct models of the Earth based on seismic data. Usually the centrifugal force is ignored, so that $\mathbf{g}_a$ is radial and all variables depend spatially only on the distance r from the geocentre. One such model is the preliminary reference Earth model (PREM) of Dziewonski and Anderson (1981); another is the model ak135 of Kennett *et al.* (1995). Both papers provide the seismic velocities and the implied ρ as functions of r, and in the case of PREM these are given as useful polynomial expressions. In their 1996 and subsequent papers, Glatzmaier and Roberts (1996a) reported the results of dynamo simulations based on PREM and on the underlying theoretical formulation of Braginsky and Roberts (1995) of thermodynamically self-consistent magnetoconvection.

Our procedure was the following: we integrated (to second order accuracy) the differential equations,

$$\frac{d\rho_a}{dr} = -\frac{\rho_a g_a}{\Phi}, \quad \frac{d}{dr}(r^2 g_a) = 4\pi G r^2 \rho_a, \tag{3.5, 3.6}$$

where the seismic parameter $\Phi = u_P^2 - 4/3 u_S^2$ was derived from the values of the seismic velocities u_P and u_S tabulated in ak135; the same density discontinuities as ak135 were assumed, namely 209 kg m^{-3} at 410 km depth, 306 kg m^{-3} at 660 km depth, 4,364 kg m^{-3} at the CMB and $\Delta\rho \equiv \rho_{SIC} - \rho_{FOC} = 565$ kg m^{-3} at the inner core boundary (ICB), "SIC" being an abbreviation for "solid inner core." This $\Delta\rho$ agrees well with that of Shearer and Masters (1991) who derived 550 ± 50 kg m^{-3}. (The discontinuities of 400 kg m^{-3} at 35 km depth and 200 kg m^{-3} at 20 km depth were also included.) Equation (3.5) is the scalar form of Eq. (3.1) with $\partial P/\partial \rho$ replaced by K/ρ, where $K = \rho\Phi$ is the incompressibility. Equation (3.6) is the Poisson equation determining g_a; here G is the gravitational constant. An arbitrary initial value for the central density ρ_c was assumed, and the equations were iterated until the correct total mass for the Earth was obtained. The values derived for ρ did not differ substantially from those tabulated in ak135. For example, we found the following: $\rho_c = 13{,}054$ kg m^{-3}; $\rho_{SIC} = 12{,}728$ kg m^{-3} and $\rho_{FOC} = 12{,}164$ kg m^{-3} at the ICB; $\rho_{FOC} = 9{,}893$ kg m^{-3} and $\rho_{mantle} = 5{,}529$ kg m^{-3} at the CMB. The corresponding ak135 values are 12,892; 12,704; 12,139; 9,915 and 5,551 kg m^{-3}, respectively. We also derived the following results, which are close to the corresponding PREM results, the latter being given in square brackets for comparison. (In the case of PREM, we did not integrate Eqs (3.5) and (3.6), but used the polynomial expressions for ρ_a given by the authors.) The central pressure is

$P_c = 363$ [361] GPa, $P_{ICB} = 328$ [329] GPa, $P_{CMB} = 135$ [136] GPa, the incompressibility $K_{ICB} = 1.29$ [1.30] GPa, $g_{ICB} = 4.38$ [4.40] m s^{-2}, $g_{CMB} = 10.68$ [10.68] m s^{-2} and at the Earth's surface $g = 9.81$ m s^{-2}; these values are close to those generally accepted. We found that the gravitational energy of the entire Earth is 2.495 [2.487] $\times 10^{33}$ J and the gravitational potential U_a is 1.116 [1.117] $\times 10^8$ m^2 s^{-2} at the geocenter. An integral of interest below is

$$\mathcal{E}_{FOC} = \int_{FOC} (U_{ICB} - U_a)\rho_a \, d^3x. \tag{3.7}$$

We found this to be 1.818 [1.817] $\times 10^{31}$ J.

3.2. Adiabatic temperature distribution

We introduced an empirical law for γ, namely that of Merkel *et al.* (2000):

$$\gamma = \gamma_0 \left(\frac{\rho_0}{\rho}\right)^q, \tag{3.8}$$

where $\gamma_0 = 1.68$ is the Grüneisen parameter for decompressed iron ($\rho_0 = 7,070$ kg m^{-3}) and $q = 0.7$. This led to $\gamma_{ICB} = 1.15$ and $\gamma_{CMB} = 1.33$. Laio *et al.* (2000) give the melting point of pure iron at the ICB pressure as $5,400$ K. The depression, ΔT_m, of the melting point through alloying is very uncertain. Williams *et al.* (1987) estimate that $\Delta T_m = 1,000$ K, while Jeanloz (1990) states that $\Delta T_m = 1,000 \pm 1,000$ K. Boehler (1996) argues that ΔT_m decreases with increasing pressure and is probably small, at most of the order of a few 100 K, at the ICB. Braginsky and Roberts (1995) worked with $\Delta T_m = 700$ K. We took $\Delta T_m = 300$ K, giving an anchor point $T_{ICB} = 5400 - \Delta T_M = 5,100$ K for the integration of Eq. (3.4). To second-order accuracy, we found that

$$T_{CMB} = 3,949 \text{ [3,954] K}, \quad \overline{T} = 4,488 \text{ [4,486] K}, \tag{3.9}$$

where $\overline{T}$ is the mass-weighted average of T in the FOC; the volume averaged T is $4,470$ [4,468] K. The adiabatic gradients at the FOC boundaries were 0.24 [0.25] K km^{-1} and 0.88 [0.85] K km^{-1}. In Section 4.2 we will encounter $\widetilde{T}$, which is the reciprocal of the mass-weighted average of $1/T_a$. For ak135 this is $4,465$ K, which differs little from $\overline{T}$.

3.3. Thermal conductivity

From our assumed value $\eta_M = 2$ m^2 s^{-1} of the magnetic diffusivity and the Wiedemann–Franz law (with a value of the Lorentz constant of 0.0194 W m s^{-2} K^{-2}) we obtained a thermal conductivity K_T at the CMB of 38 W m^{-1} K^{-1}. Stacey and Anderson (2001) supplement this electronic part with a lattice contribution, leading them to 46 W m^{-1} K^{-1}. We dealt with this point by multiplying our K_T everywhere by the appropriate constant factor, giving, for example, $K_T = 57$ W m^{-1} K^{-1} at the ICB. Having derived in this way K_T as a function of r, we could determine the adiabatic heat flux, which we found to be 0.0138 [0.0141] W m^{-2} at the ICB and 0.0386 [0.0373] W m^{-2} at the CMB. The corresponding net adiabatic heat flows are

$$\mathcal{H}_{ICB} = 0.26 \text{ [0.26] TW}, \quad \mathcal{H}_{CMB} = 5.87 \text{ [5.68] TW}. \tag{3.10}$$

The expenditure of entropy by the adiabatic gradient in the FOC is

$$\Sigma = \int_{FOC} K_M (\nabla T_a / T_a)^2 \, dV \approx 172 \text{ [167] MW K}^{-1}. \tag{3.11}$$

This estimate omits the contributions to Σ from boundary layers, which are assumed to be negligible.

3.4. Specific and latent heats; contraction on solidification

We derived estimates of other thermal properties in the FOC, as functions of r, by the following process. To compute c_p and α, we adopted the approach of Braginsky and Roberts (1995), who took c_v, the specific heat at constant volume, to be $670A^{\mathrm{Fe}}/\overline{A}\,\mathrm{J\,kg^{-1}\,K^{-1}}$, where $A^{\mathrm{Fe}} = 55.85$ is the atomic weight of iron and $\overline{A} = 48.1$ is the mean atomic weight of the core (see, e.g. Stacey, 1992). From this they derived c_p from Eq. (E7) of Braginsky and Roberts (1995):

$$c_p = \frac{c_v}{1 - \gamma^2 c_v T/u_P^2}. \tag{3.12}$$

They then obtained $\alpha = \gamma c_p/u_P^2$.

We found that, at the ICB, $c_p = 819\,[820]\,\mathrm{J\,kg^{-1}K^{-1}}$ and $\alpha = 0.89\,[0.90] \times 10^{-5}\,\mathrm{K^{-1}}$; at the CMB, $c_p = 850\,[848]\,\mathrm{J\,kg^{-1}\,K^{-1}}$ and $\alpha = 1.77\,[1.70] \times 10^{-5}\,\mathrm{K^{-1}}$. Having found c_p, we can transform the thermal conductivity K_T into a thermal diffusivity $\kappa_M = K_T/\rho c_p$, and we find that $(\kappa_M)_{\mathrm{ICB}} = 5.7 \times 10^{-6}\,\mathrm{m^2\,s^{-1}}$ and $(\kappa_M)_{\mathrm{CMB}} = 5.4 \times 10^{-6}\,\mathrm{m^2\,s^{-1}}$.

We follow Anderson and Duba (1997) by supposing that the change in specific volume on freezing is $2.0 \times 10^{-6}\,\mathrm{m^3\,kg^{-1}}$, corresponding to a contribution of $\Delta^s\rho = 310\,\mathrm{kg\,m^{-3}}$ to $\Delta\rho$, the remainder, $\Delta^\xi\rho = 255\,\mathrm{kg\,m^{-3}}$, being due to the rejection of light constituent on freezing core mix. Braginsky and Roberts (1995) divided the PREM value of $\Delta\rho = 600\,\mathrm{kg\,m^{-3}}$ differently, taking $\Delta^s\rho = 100\,\mathrm{kg\,m^{-3}}$ and $\Delta^\xi\rho = 500\,\mathrm{kg\,m^{-3}}$. This is the primary reason why our estimates below differ from theirs.

It may be worth pointing out that the smaller value of $\Delta^\xi\rho$ assumed by Anderson and Duba (1997) as compared with Braginsky and Roberts (1995) diminishes the potency of compositional buoyancy. Thermal buoyancy becomes more significant and plausibly the character of the geodynamo does not undergo as marked a change as the SIC comes into existence. If a similar situation arises in the core of Venus, one may speculate that the possible absence of an inner core in that planet may not explain why it does not create a magnetic field. (For an alternative point of view, see Stevenson *et al.*, 1983.)

We used the expression for the latent heat of crystallization, L, given by Anderson and Duba (1997):

$$L = \left[\frac{K}{2(\gamma - (1/3))(1 + \gamma^2\alpha T)}\right]_{\mathrm{ICB}} \Delta^s V. \tag{3.13}$$

This gives $L = 1.56 \times 10^6\,\mathrm{J\,kg^{-1}}$, instead of the value of $10^6\,\mathrm{J\,kg^{-1}}$ of Braginsky and Roberts (1995), which they drew from Gubbins *et al.* (1979). Labrosse *et al.* (1997, 2001) took $L = 0.625 \times 10^6\,\mathrm{J\,kg^{-1}}$, which they derived from Poirier and Shankland (1993) using $\Delta^s V = 0.93\,\mathrm{m^3\,kg^{-1}}$. Laio *et al.* (2000) obtained $L = 0.7 \times 10^6\,\mathrm{J\,kg^{-1}}$.

3.5. Light constituent

Braginsky and Roberts (1995) argued for a value of about 0.16 for the mass fraction ξ_a of light constituent in the well-mixed adiabatic state but later research (e.g. Anderson and Duba, 1997) suggests that 10% is a better estimate; Loper (1991) took 9%. In what follows, we take $\xi_a = 0.095$. Braginsky and Roberts (1995) related ξ_a to $\Delta\xi$, where $\Delta\xi \equiv \xi_{\mathrm{FOC}} - \xi_{\mathrm{ICB}}$

is the discontinuity in ξ at the ICB, by a "rejection factor," $r_{FS} = \Delta\xi/\xi_a$ and showed that, to a good approximation,

$$r_{FS} \equiv \frac{\Delta\xi}{\xi_a} = \frac{\Delta^\xi \rho/\rho_{SIC}}{1 - \rho/\rho_H}, \tag{3.14}$$

where ρ/ρ_H is the ratio of ρ to the density that pure iron would have under the same conditions. They presented an argument that led to the estimate $\rho_H/\rho = 1.1052$, which we adopt for the SIC. This leads from $\xi_a = 0.095$ to $\Delta\xi = 0.020$. Note that this value of $\Delta\xi$ is significantly less than the estimate of 0.06 given in Braginsky and Roberts (1995). The change is due partly to the revised partition of $\Delta\rho$ and partly due to the reduced estimate of ξ_a, both changes decreasing the estimated $\Delta\xi$.

A further significant parameter is the compositional expansion coefficient, α^ξ, which at the bottom of the FOC is

$$\alpha^\xi \equiv \frac{1}{\rho}\left(\frac{\partial\rho}{\partial\xi}\right)_{P,S} = \frac{\Delta^\xi \rho}{\rho_{SIC}\Delta\xi} \tag{3.15}$$

here, as in Eq. (3.14) and in Eq. (3.18) below, ρ_{SIC} refers to the top of the SIC. We suppose that the value $\alpha^\xi = 1.0$ obtained from Eq. (3.15) holds throughout the FOC; this may be compared with $\alpha^\xi = 0.6$, as used by Braginsky and Roberts (1995).

A quantity that is significant when discussing the advance of the ICB through freezing is

$$\Delta_2 = \Delta_{ma} + \Delta_{m\xi}, \tag{3.16}$$

where

$$\Delta_{ma} = \left(\frac{\gamma gr}{u_P^2}\right)_{ICB}\left(\frac{\partial T_m/\partial P}{\partial T_a/\partial P} - 1\right), \tag{3.17}$$

$$\Delta_{m\xi} = -\frac{3V_{SIC}\rho_{SIC}\Delta\xi}{\mathcal{M}_{FOC}T}\left[\frac{h^\xi}{c_p} + \frac{\partial T_m}{\partial\xi}\right]. \tag{3.18}$$

Here $V_{SIC} \approx 1.765 \times 10^{20}$ m^3 is the volume of the SIC and $\mathcal{M}_{FOC} \approx 1.8412 \times 10^{24}$ kg is the mass of the FOC. We took the heat of reaction, h^ξ, to be -1.6×10^7 J kg^{-1}, a value quoted by Gubbins (1977) for FeS; Braginsky and Roberts (1995) took $h^\xi = -5 \times 10^6$ J kg^{-1} for this rather uncertain parameter. We use the crude estimate $\partial T_m/\partial\xi = \Delta T_m/\Delta\xi$. Recalling that we are assuming $\Delta T_m = 300$ K, we find that $\Delta_{m\xi} = 0.014$. Estimating the melting point gradient $\partial T_m/\partial P$ by using Lindemann's law, which Braginsky and Roberts (1995) showed reduces approximately to

$$\frac{\partial T_m/\partial P}{\partial T_a/\partial P} - 1 = 1 - \frac{2}{3\gamma}, \tag{3.19}$$

we obtain $\Delta_{ma} = 0.024$, giving $\Delta_2 = 0.038$ [0.038] in total. Braginsky and Roberts obtained $\Delta_{ma} = 0.03$, $\Delta_{m\xi} = 0.02$ and $\Delta_2 = 0.05$ in total.

4. Energy and entropy

4.1. Energy balance

The energy balance of the FOC is expressed by

$$\mathcal{Q}_{CMB} = \mathcal{Q}_{ICB} + \mathcal{Q}^L + \mathcal{Q}^S + \mathcal{Q}^G + \mathcal{Q}^R, \tag{4.1}$$

where $\mathcal{Q}_{ICB}$ and $\mathcal{Q}_{CMB}$ are the heat flows in and out of the boundaries, and $\mathcal{Q}^L$ is the latent heat released at the SIC into the FOC. The remaining terms are volumetric sources within the FOC: $\mathcal{Q}^R$ is the rate of supply of the radioactive sources; $\mathcal{Q}^S$ is rate of diminution of internal energy through cooling; and $\mathcal{Q}^G$ is the gravitational energy released through the emission of light constituent during freezing at the ICB, this constituent being mixed thoroughly by convection throughout the FOC. In writing equations such as (4.1), we assume that the quantities concerned are averaged over a few convective overturning times, that is, of the order of a thousand years, say.

How are motions in the Earth's core created? The obvious answer is that radioactive elements dissolved in the core, heat the fluid and provide the buoyancy that drives thermal convection. The view of the geochemists has however hardened against this idea, and it is now widely stated that there is no significant radioactivity in the core; see, for example, Stacey (1992). Recently this view has been challenged (see Section 6), and we shall suppose here that radioactivity supplies energy to the core at a rate, $\mathcal{Q}^R$, that is not necessarily zero. We shall, nevertheless, at first accept the traditional view and set $\mathcal{Q}^R = 0$. This does not rule out thermal convection as a source of motion. If there are no radioactive sources in the core, it must be cooling, and when bodies of fluid are cooled sufficiently rapidly they convect.

The terms $\mathcal{Q}^L$, $\mathcal{Q}^S$ and $\mathcal{Q}^G$ are all proportional to the rate of advance of the ICB. We incorporate this effect through a parameter λ which crudely represents the age τ_{SIC} of the inner core, normalized to 1.2 Gyr, for example, to obtain these sources for $\tau_{SIC} = 0.6$ Gyr, we should set $\lambda = 1/2$ in Eqs (4.3), (4.6) and (4.9). More precisely, we suppose that $\mathcal{M}_{SIC} = \mathcal{M}_{SIC}/(\lambda \times 1.2 \, \text{Gyr})$, where λ is constant. The radioactive source is also time dependent, decreasing exponentially with time, according to the half-life of the radioactive element concerned; though easily incorporated, we do not include that effect here.

Because of the convection, the actual heat, $\mathcal{Q}_{ICB}$ and $\mathcal{Q}_{CMB}$, entering and leaving the fluid core may be expected to differ from $\mathcal{H}_{ICB}$ and $\mathcal{H}_{CMB}$, perhaps substantially. If we take $\kappa_M = 6 \times 10^{-6} \, \text{m}^2 \, \text{s}^{-1}$ in the SIC, we find that its thermal time constant r_{ICB}^2/κ_M is about 6 Gyr, which suggests that T is "frozen-in" to the SIC, and that ∇T is of the order of the melting point gradient, which is less than the adiabatic gradient in the overlying FOC. If this is the case, $\mathcal{Q}_{ICB}$ exceeds $\mathcal{H}_{ICB}$, though not by a large factor. Since, however, the following arguments are insensitive to the value we assume for $\mathcal{Q}_{ICB}$, we shall for simplicity assume that

$$\mathcal{Q}_{ICB} = \mathcal{H}_{ICB}. \tag{4.2}$$

The rate, $\mathcal{Q}^L$, at which latent heat is released at the ICB is $L\dot{\mathcal{M}}_{SIC}$, which for our chosen material properties is

$$\mathcal{Q}^L = 4.0/\lambda \, \text{TW}. \tag{4.3}$$

The sources $\mathcal{Q}^S$ and $\mathcal{Q}^G$ are, in general, harder to estimate. Because α and $\Delta^s \rho$ are nonzero, the Earth contracts as it cools, and the CMB moves systematically inwards, releasing gravitational energy as it does so. The inward motion is associated with "$P \, dV$ work" that increases the internal energy of the FOC. According to Loper (1991), the energy associated with contraction is surprisingly large,[1] although Gubbins *et al.* (1979) argue that it is not useful work for driving the dynamo. Matters such as these greatly complicate the theoretical analysis of the energy and entropy budgets, and in the interests of simplicity we eliminate them. This is easily done by assuming, as Braginsky and Roberts (1995) did, that the radius, r_{CMB}, of the CMB is constant. The release of gravitational energy then comes about because $\Delta^\xi \rho$ is nonzero; the fractionation of constituents at the ICB causes the buoyant light material released to move upwards and the heavy constituent to move downwards, that is, the Earth is continually becoming more centrally condensed.

Considering $\mathcal{Q}^S$ first, we have

$$\mathcal{Q}^S = -\int_{FOC} \dot{S}_a T_a \rho_a \, d^3x = -\dot{S}_a \int_{FOC} T_a \rho_a \, d^3x = -\dot{S}_a \overline{T} \mathcal{M}_{FOC}. \tag{4.4}$$

Braginsky and Roberts (1995) showed that, to a good approximation,

$$\dot{S}_a = -\Delta_2 c_p \dot{r}_{ICB}/r_{ICB}, \tag{4.5}$$

where Δ_2 is the constant defined in Eqs (3.16)–(3.18) above. Assuming the value $\Delta_2 = 0.038$ obtained there and substituting $\dot{r}_{ICB} = \mathcal{M}_{SIC}/4\pi r_{ICB}^2 \rho_{ICB}$, we find that Eqs (4.4) and (4.5) give

$$\mathcal{Q}^S = 2.3/\lambda \, \text{TW}. \tag{4.6}$$

Gravitational energy is released into the FOC at the rate

$$\mathcal{Q}^G = -\dot{\xi}_a \int_{FOC} (\mu_a - \mu_{ICB})\rho_a \, d^3x, \tag{4.7}$$

where μ is the chemical potential. Since α^ξ is uniform in the FOC, by assumption, we may replace Eq. (4.7) by

$$\mathcal{Q}^G = \dot{\xi}_a \alpha^\xi \int_{FOC} (U_{ICB} - U_a)\rho_a \, d^3x = \dot{\xi}_a \alpha^\xi \mathcal{E}_{FOC}, \tag{4.8}$$

where Eqs (3.7) and (3.15) provide $\mathcal{E}_{FOC}$ and α^ξ. Since $\dot{\xi}_a = 4\pi r_{ICB}^2 \dot{r}_{ICB}^2 \rho_{SIC} \Delta \xi / \mathcal{M}_{FOC}$ (see Braginsky and Roberts, 1995), it follows that

$$\mathcal{Q}^G = 0.5/\lambda \, \text{TW}. \tag{4.9}$$

Collecting together our results, we have

$$\mathcal{Q}_{CMB} = \frac{6.8 \, \text{TW}}{\lambda} + 0.3 \, \text{TW} + \mathcal{Q}^R. \tag{4.10}$$

The material properties assumed by Braginsky and Roberts (1995) imply values of the constants featuring in Eqs (4.3), (4.6) and (4.9) that are substantially different from ours; they were 2.5, 3.1 and 1.3, respectively. But these values give almost the same result as Eq. (4.10); the 6.8 appearing in Eq. (4.10) was replaced by 6.9, so that $\mathcal{Q}^R = 0$ and $\mathcal{Q}_{CMB} = 7.2 \, \text{TW}$, which were the values assumed by Glatzmaier and Roberts (1996a) in their simulation, imply that $\lambda = 1$, that is, $\tau_{SIC} = 1.2 \, \text{Gyr}$. Not surprisingly, though perhaps comfortingly, the rate of advance of the ICB in the simulation had already indicated that $\tau_{SIC} \approx 1.2 \, \text{Gyr}$. Despite the similarity of the two estimates for $\mathcal{Q}_{CMB}$ (7.1 and 7.2 TW), the gross thermodynamics of the two models are rather different, as is indicated below.

If $\tau_{SIC} = 4 \, \text{Gyr}$ and $\mathcal{Q}^R = 0$, Eq. (4.10) gives $\mathcal{Q}_{CMB} = 2.3 \, \text{TW}$, which is less than a half of the adiabatic heat flux $\mathcal{H}_{CMB}$. As Loper (1978) pointed out, the superadiabatic heat flow at the CMB, which is

$$\mathcal{Q}_{sa} \equiv \mathcal{Q}_{CMB} - \mathcal{H}_{CMB} = \left(\frac{6.8}{\lambda} - 5.6\right) \text{TW}, \tag{4.11}$$

may in principle be either positive or negative. It is not impossible, as Glatzmaier and Roberts (1997) show through a concrete example, for a dynamo with a subadiabatic $\mathcal{Q}_{CMB}$ to function, but it is unlikely that such a dynamo would be vigorous enough to undergo polarity reversals. This illustrates the difficulty in supposing that the solid inner core is an old feature of the Earth.

The rate of working of the buoyancy forces does not feature in Eq. (4.1), since the kinetic and magnetic energies buoyancy creates are returned to the core as heat through ohmic and viscous dissipation.

4.2. Entropy balance

The energy balance does not suffice to describe the gross state of the core completely. The entropy balance, which depends on the available dissipation, adds a further constraint. This was first recognized by Braginsky (1964). The subject was advanced by Backus (1975), Hewitt *et al.* (1975), Gubbins *et al.* (1979), Braginsky and Roberts (1995) and others.

The entropy balance of the FOC is expressed by

$$\frac{\mathcal{Q}_{CMB}}{T_{CMB}} = \frac{\mathcal{Q}_{ICB} + \mathcal{Q}^L}{T_{ICB}} + \Sigma + \frac{\mathcal{Q}^D}{T_D} + \frac{\mathcal{Q}^S + \mathcal{Q}^R}{\overline{T}}, \qquad (4.12)$$

where $\mathcal{Q}_{ICB}/T_{ICB}$ and $\mathcal{Q}_{CMB}/T_{CMB}$ are the entropy flows in and out of the boundaries, and $\mathcal{Q}^L/T_{ICB}$ is the entropy released at the SIC through freezing. The remaining terms are volumetric sources within the FOC and are discussed more fully below.

Entropy is produced volumetrically through the diffusion of field, vorticity, heat and composition. Of these, the most significant is Σ, which arises from heat conduction down the adiabatic gradient and which is evaluated in Eq. (3.11). In calculating the remaining diffusive contributions, we adopt the two-scale strategy of Section 1. Because Re, Pe and Mp are large, the entropy production per unit volume by the macroscale is dominated by $\overline{\sigma} = q^{\overline{J}}/T_a$, where $q^{\overline{J}} = \mu_0 \eta_M \overline{\mathbf{J}}^2$ is the macroscopic dissipation rate, see Section 1. The entropy production by thermal and compositional diffusion is dominated by turbulent diffusion, through the fluxes $\mathbf{I}^S \equiv \overline{S'\mathbf{V}'}$ and $\mathbf{I}^{\xi} \equiv \overline{\xi'\mathbf{V}'}$ where, according to the Reynolds analogy, $\mathbf{I}^S = -\rho_a \kappa_T \nabla \overline{S}$ and $\mathbf{I}^{\xi} = -\rho_a \kappa_T \nabla \overline{\xi}$; see Section 1. Braginsky and Roberts (1995) showed in Section 4.2 and Appendix C of their paper that the resulting mean entropy production could be written as $\sigma' = -\mathbf{g}_a \cdot (\alpha^S \mathbf{I}^S + \alpha^{\xi} \mathbf{I}^{\xi})/T_a$, that is, as

$$\sigma' = \rho_a \kappa_T \mathbf{g}_a \cdot (\alpha^S \nabla \overline{S} + \alpha^{\xi} \nabla \overline{\xi})/T_a, \qquad (4.13)$$

an expression that they related to the average rate at which buoyancy feeds energy into the microscale to maintain the turbulence. This form of σ' is useful in numerical computations, and was adopted by Glatzmaier and Roberts (1996a), who used it to obtain the estimate $\mathcal{Q}^{J'} = 0.1$ TW mentioned in Section 2.1 (and there criticized as being too small).

Braginsky and Roberts (1995) also showed that the mean entropy production by the microscale could be written as $\sigma' = q^{J'}/T_a$ per unit volume, where $q^{J'} = \mu_0 \eta_M \overline{(\mathbf{J}')^2}$ is the average microscopic ohmic dissipation, see Section 1. At first sight, this result may seem surprising. One way of rationalizing it is to imagine oneself moving forward in time along a hypothetical sequence at the end point of which, in the remote future, computer technology has developed to the point where $\mathbf{V}$, $\mathbf{B}$, S and ξ are fully resolved, including the parts we have called $\mathbf{V}'$, $\mathbf{B}'$, S' and ξ'. The length and time scales become increasingly well resolved along this sequence, the turbulent viscosity ν_T and the turbulent diffusivity κ_T of heat and composition become increasingly irrelevent, and finally only the molecular diffusivities ν_M, κ_M and D_M act. Because these are small compared with η_M, their contributions to the entropy production will then be negligible compared with the ohmic rate of entropy production. And this, because it now is calculated from electric currents of all length scales, has increased to what, according to the two-scale ansatz, has been written as $(q^J + q^{J'})/T_a = q^J/T_a$. In short, the entropy production is the same at one end of our hypothetical sequence as the other, but it is partitioned differently. Initially, it is Eq. (4.13) but finally it is what we have called $q^{J'}/T_a$.

In recognition of the fact that Joule losses dominate the dissipative processes, we write $q^J + q^{J'}$ as q^D rather than q^J. Integrating q^D over the FOC, we obtain a net dissipation rate of

$$Q^D = \int_{FOC} q^D \, d^3x, \tag{4.14}$$

and in Eq. (4.12) we have, in the interests of simplicity, written the net entropy production rate as

$$\frac{Q^D}{T_D} = \int_{FOC} \frac{q^D}{T_a} \, d^3x. \tag{4.15}$$

The estimates of Q^J made in Section 2 contain all the information about Q^D that we require.

It is clear that $T_{CMB} < T_D < T_{ICB}$ and, since the superadiabatic gradient is almost certainly largest near the bottom of the FOC (leading to more active convection, field generation and associated dissipation), it is plausible that $T_D > \overline{T}$. The Glatzmaier and Roberts simulations tend to show strong activity within and deep in the tangent cylinder and, although currents generated in these regions leak out into the exterior of the tangent cylinder, the main Joule heat losses appear to be within it, again suggesting that $T_D > \overline{T}$. It is possible in principle to estimate T_D from a dynamo simulation, but in practice the hyperdiffusion demanded for numerical reasons makes such estimates unreliable.

The entropy source associated with the radiogenic heat sources in the FOC may be similarly written as

$$\frac{Q^R}{T_R} = \int_{FOC} \frac{q^R}{T_a} \, d^3x. \tag{4.16}$$

Since these sources are thoroughly mixed by the convection, the rate q^R/ρ_a at which they inject energy is presumably uniform over the FOC, so that $T_R = \widetilde{T}$. Here $1/\widetilde{T}$ is the mass weighted average of $1/T_a$, which was evaluated in Section 3.2. In what follows we shall, for simplicity, take $T_R = \overline{T}$.

The entropy source associated with the loss of internal energy through the cooling of the Earth is

$$\int_{FOC} \dot{S}_a \rho_a \, d^3x = \mathcal{M}_{FOC} \dot{S}_a = -\frac{Q^S}{\overline{T}}, \tag{4.17}$$

by Eq. (4.4).

According to Eqs (4.1), (4.2) and (4.12), we now have

$$Q^D = Q^G + Q^H, \tag{4.18}$$

where Q^H is the part of Q^D available from heat sources:

$$Q^H = Q_{CMB}\left(\frac{T_D}{T_{CMB}} - 1\right) + (\mathcal{H}_{ICB} + Q^L)\left(1 - \frac{T_D}{T_{ICB}}\right)$$
$$+ (Q^S + Q^R)\left(1 - \frac{T_D}{T}\right) - \Sigma T_D. \tag{4.19}$$

Thus, as anticipated by Braginsky (1964), all the gravitational energy released is available for dissipation but only a part, proportional to $\Delta T / T$ for some ΔT, of each thermal contribution to Q^D can be used; in the case of Σ, this is evident from its definition (3.11).

This emasculation of the heat sources is reminiscent of the heat engines of classical thermodynamics, and in particular of the Carnot cycle. In such cases, Q^D might, since it relates to the creation of the observed geomagnetic field, be called "useful work," but such a description is inappropriate when this energy is unavailable outside the system but is re-deposited within it. The term "available dissipation" may be preferable.

In Section 4.3 we shall make use of Eqs (4.1) and (4.2) to write Eqs (4.18) and (4.19) as

$$Q^D = \frac{T_D}{T_{CMB}}\left[(\mathcal{H}_{ICB} + Q^L)\left(1 - \frac{T_{CMB}}{T_{ICB}}\right) + (Q^S + Q^R)\left(1 - \frac{T_{CMB}}{T}\right) + Q^G - \Sigma T_{CMB}\right]. \tag{4.20}$$

Using the constants we have derived above, we may write this as

$$Q^D = \frac{T_D}{T_{CMB}}\left(\frac{1.69\,\text{TW}}{\lambda} - 0.62\,\text{TW} + 0.12 Q^R\right). \tag{4.21}$$

Let us suppose (until Section 6) that $Q^R = 0$. Then it is clear from Eq. (4.21) that, since Q^D must be positive, $\lambda < 2.7$, that is, $\tau_{SIC} < 3.3\,\text{Gyr}$. This crude estimate is refined if we place Q^D in the range (2.19). We examine two possibilities, case (a) modeling a dynamo where the Joule heat loss is uniform over the core and case (b) modeling a dynamo where the dissipation occurs preferentially nearer the ICB:

$$\text{(a)} \quad T_D = \overline{T} = 4,488\,\text{K}, \quad \text{(b)} \quad T_D = \tfrac{1}{2}(\overline{T} + T_{ICB}) = 4,794\,\text{K}.$$

We then find that, corresponding to $1\,\text{TW} < Q^D < 2\,\text{TW}$, we have

$$\text{(a)} \quad 6.3\,\text{TW} < Q_{CMB} < 9.9\,\text{TW}, \quad 1.34\,\text{Gyr} > \tau_{SIC} > 0.85\,\text{Gyr}, \tag{4.22a}$$
$$\text{(b)} \quad 6.1\,\text{TW} < Q_{CMB} < 9.4\,\text{TW}, \quad 1.40\,\text{Gyr} > \tau_{SIC} > 0.89\,\text{Gyr}. \tag{4.22b}$$

Let us define the "critical dynamo" as one in which the Nusselt number, $Nu \equiv Q_{CMB}/\mathcal{H}_{CMB}$, is 1. According to Eqs (4.10) and (4.21), this corresponds to $\tau_{SIC} = 1.46\,\text{Gyr}$ (and $Q^D = 0.87\,\text{TW}$). Subcritical dynamo action is not impossible, as was established by a model of Glatzmaier and Roberts (1997), but that dynamo could only produce a field small compared with the Earth's, one that gave no indications of reversing. While other factors are involved, such as the way the flow of heat Q_{CMB} out of the core is partitioned over the CMB (Glatzmaier *et al.*, 1999), numerical simulations suggest that the larger the value of Nu the greater the frequency of reversals. The upper bounds in Eqs (4.22a) and (4.22b) correspond to $Nu \approx 1.7$. This may be compared with Nu for the eight models studied by (Glatzmaier *et al.*, 1999), which was only about 1.3 even though seven of the models reversed (some frequently). Probably, the upper bounds on Q_{CMB} in Eqs (4.22a) and (4.22b) correspond to dynamos that reverse fairly frequently, while the lower bounds correspond to slightly super-critical dynamos that are likely to be weak and nonreversing. The thermodynamic arguments we present here suggest that the distinction between these two types of behavior is better made on the basis of Q^D than on the basis of Nu.

A striking feature of Eq (4.22) is the youth of the inner core, even when the upper bounds for τ_{SIC} are assumed. This raises the question of how the geodynamo functioned early in the Earth's history, when the sources of latent heat and light component at the ICB were absent. We discuss this briefly in Section 4.4.

4.3. Variants

This subsection explores the effect of changing each of the four most uncertain parameters: the contraction on solidification, $\Delta^s V$; the depression of the melting point through alloying, T_m; the mass fraction of the light component, ξ_a and the heat of reaction h^ξ. Three of the values

$$\Delta^s V = 2 \times 10^{-6}\,\mathrm{m}^3\,\mathrm{kg}^{-1}, \quad T_m = 300\,\mathrm{K}, \quad \xi_a = 0.095, \quad h^\xi = -1.6 \times 10^7\,\mathrm{J\,kg}^{-1} \tag{4.23}$$

are held fixed as the fourth is changed. The results are shown in Tables 1–4, where it is supposed throughout that the melting point of pure iron is 5,400 K. The values of Q^L, Q^S and Q^G assume that the age of the inner core is $\tau_{SIC} = 1.2$ Gyr; if instead it is 1.2λ Gyr, the values should be divided by λ. We also write the energy and entropy balances as

$$Q_{CMB} = \frac{E_\lambda}{\lambda} + \mathcal{H}_{ICB} + Q^R, \tag{4.24}$$

$$Q^D = \frac{T_D}{T_{CMB}} \left(\frac{D_\lambda}{\lambda} - D_0 + 0.12 Q^R \right). \tag{4.25}$$

We have here given $1 - T_{CMB}/\overline{T}$ the value 0.12 that it has in every case considered here.

It is clear from Table 1 that, as anticipated in the Section 3.4, many results are sensitive to $\Delta^s V$. Table 2 shows that T_{ICB}, T_{CMB}, and $\overline{T}$ are very dependent on T_m; also $\mathcal{H}_{ICB}$, $\mathcal{H}_{CMB}$ and D_0 vary by about a factor of 2 across the range shown but other quantities, including Q^L, Q^S, E_λ and D_λ, are less affected. The constants appearing in Eqs (4.24) and (4.25) are also fairly insensitive to ξ_a and h^ξ, as is shown in Tables 3 and 4.

In Section 4.2, we explored relations (4.10) and (4.21) for our "standard model" (4.23) in three ways. First, we used them to set a crude upper limit on the age of the inner core, τ_{SIC}; second we used the upper and lower bounds in Eq. (2.19) to limit τ_{SIC} and to calculate the corresponding Q_{CMB}; third, we supposed that the Nusselt number, $Nu \equiv Q_{CMB}/\mathcal{H}_{CMB}$ was unity and determined τ_{SIC} for that "critical state," together with the corresponding value of Q^D. In all cases Q^R was assumed to be zero. Since the differences between Eqs (4.22a) and (4.22b) were not substantial, we confine attention here to case (a), in which $T_D = \overline{T}$. Tables 1–4 show the corresponding results for all the models considered. The crude limit is denoted by τ_{SIC}^{crude}; values for the lower limit in Eqs (4.22a) and (4.22b) corresponding to $Q^D = 1$ TW carry a superfix 1 and those for the upper limit for $Q^D = 2$ TW carry the superfix 2; values denoted by τ_{SIC}^c and Q_c^D refer to the critical case $Nu = 1$. Perhaps the most striking fact that emerges from the tabulated values is by how little they vary, even for the different values of $\Delta^s V$. In particular, all the upper bounds τ_{SIC}^1 of τ_{SIC} are considerably less than the known age of the geomagnetic field (Kono and Tanaka, 1995).

Table 1 The effect of varying $\Delta^s V$

$10^6 \Delta^s V$ (m³ kg⁻¹)	0	1	2	3	3.65
$\Delta^s \rho$ (kg m⁻³)	0	155	310	464	565
$\Delta^\xi \rho$ (kg m⁻³)	565	410	255	101	0
r_{FS}	0.47	0.34	0.21	0.083	0
$\Delta\xi$	0.044	0.032	0.020	0.008	0
$100\Delta_2$	5.53	4.68	3.83	2.98	2.43
L (J kg⁻¹)	0	0.78	1.56	2.33	2.84
Q^L (TW)	0	2.00	4.00	6.00	7.30
Q^S (TW)	3.33	2.82	2.31	1.80	1.47
Q^G (TW)	1.13	0.82	0.51	0.20	0
E_λ (TW)	4.45	5.63	6.82	7.99	8.76
D_λ (TW)	1.53	1.61	1.69	1.77	1.82
τ_{SIC}^{crude} (Gyr)	2.9	3.1	3.3	3.4	3.5
Q_{CMB}^1 (TW)	4.6	5.5	6.3	7.1	7.5
τ_{SIC}^1 (Gyr)	1.22	1.29	1.34	1.41	1.45
Q_{CMB}^2 (TW)	7.2	8.6	9.9	11.0	11.7
τ_{SIC}^2 (Gyr)	0.77	0.81	0.85	0.89	0.92
Q_c^D (TW)	1.48	1.11	0.87	0.70	0.62
τ_{SIC}^c (Gyr)	0.95	1.20	1.46	1.71	1.87

Table 2 The effect of varying T_m

T_m (K)	0	300	700	1,000	1,500
T_{ICB} (K)	5,400	5,100	4,700	4,400	3,900
T_{CMB} (K)	4,182	3,950	3,640	3,408	3,021
$\bar{T}$ (K)	4,753	4,488	4,136	3,872	3,432
$\mathcal{H}_{ICB}$ (TW)	0.29	0.26	0.22	0.19	0.15
$\mathcal{H}_{CMB}$ (TW)	6.58	5.87	4.99	4.37	3.43
Σ (GW K⁻¹)	0.183	0.172	0.159	0.149	0.132
L (J kg⁻¹)	1.55	1.56	1.57	1.57	1.58
$100\Delta_2$	1.83	1.72	1.59	1.49	1.32
Q^L (TW)	3.99	4.00	4.02	4.04	4.06
Q^S (TW)	2.28	2.31	2.34	2.37	2.41
E_λ (TW)	6.78	6.82	6.87	6.91	6.98
D_0 (TW)	0.70	0.62	0.53	0.46	0.36
D_λ (TW)	1.68	1.69	1.70	1.70	1.71
τ_{SIC}^{crude} (Gyr)	2.9	3.3	3.8	4.4	5.6
Q_{CMB}^1 (TW)	6.6	6.3	5.9	5.6	5.2
τ_{SIC}^1 (Gyr)	1.28	1.34	1.44	1.52	1.65
Q_{CMB}^2 (TW)	10.2	9.9	9.5	9.2	8.8
τ_{SIC}^2 (Gyr)	0.82	0.85	0.89	0.92	0.97
Q_c^D (TW)	0.98	0.87	0.74	0.64	0.50
τ_{SIC}^c (Gyr)	1.29	1.46	1.73	1.98	2.55

4.4. The early Earth

It was mentioned in Section 4.1 that the Glatzmaier and Roberts (1996a) model was consistent with an inner core that is only 1.2 Gyr old. Other research has led to similar estimates of τ_{SIC}. Stevenson *et al.* (1983) estimated that $\tau_{SIC} \lesssim 2.3$ Gyr, Labrosse *et al.* (1997) that

Table 3 The effect of varying ξ_a

ξ_a	0.03	0.095	0.12	0.16	0.2
$\Delta\xi$	0.011	0.020	0.025	0.034	0.042
α^ξ	1.90	1.00	0.79	0.59	0.48
$100\Delta_2$	3.26	3.83	4.15	4.65	5.16
$\mathcal{Q}^S$ (TW)	1.96	2.31	2.50	2.80	3.11
E_λ (TW)	6.47	6.82	7.00	7.31	7.61
D_λ (TW)	1.65	1.69	1.71	1.75	1.78
τ_{SIC}^{crude} (Gyr)	3.1	3.3	3.3	3.4	3.4
$\mathcal{Q}_{CMB}^1$ (TW)	6.1	6.3	6.4	6.5	6.7
τ_{SIC}^1 (Gyr)	1.30	1.34	1.37	1.40	1.42
$\mathcal{Q}_{CMB}^2$ (TW)	9.5	9.9	10.0	10.2	10.4
τ_{SIC}^2 (Gyr)	0.82	0.85	0.86	0.88	0.90
$\mathcal{Q}_c^D$ (TW)	0.94	0.87	0.85	0.82	0.79
τ_{SIC}^c (Gyr)	1.35	1.46	1.50	1.56	1.63

Table 4 The effect of varying h^ξ

$-10^7 h^\xi$ $(J\,kg^{-1})$	0	1	1.6	2	5
$100\Delta_2$	2.63	3.38	3.83	4.13	6.39
$\mathcal{Q}^S$ (TW)	1.58	2.04	2.31	2.49	3.85
E_λ (TW)	6.09	6.54	6.82	7.00	8.35
D_λ (TW)	1.60	1.65	1.69	1.71	1.87
τ_{SIC}^{crude} (Gyr)	3.1	3.2	3.3	3.3	3.6
$\mathcal{Q}_{CMB}^1$ (TW)	6.0	6.2	6.3	6.4	7.0
τ_{SIC}^1 (Gyr)	1.28	1.32	1.34	1.37	1.49
$\mathcal{Q}_{CMB}^2$ (TW)	9.3	9.7	9.9	10.0	10.9
τ_{SIC}^2 (Gyr)	0.81	0.83	0.85	0.86	0.94
$\mathcal{Q}_c^D$ (TW)	0.97	0.91	0.87	0.85	0.72
τ_{SIC}^c (Gyr)	1.30	1.40	1.46	1.50	1.78

$\tau_{SIC} \lesssim 1.7$ Gyr and Buffett *et al.* (1996) that $\tau_{SIC} \approx 2.8$ Gyr. But it is known that the geomagnetic field has existed, with an amplitude within a factor of 4 of its present strength, for at least 3.5 Gyr; see Kono and Tanaka (1995). It is of interest to consider how the estimates of Section 4.2 have to be modified prior to the formation of the inner core.

In looking at this question in slightly greater depth, we make some simplifications. The present SIC is small compared with the FOC, both in terms of mass and volume, and this was even more true in the past. We, therefore, trust that we are not making a serious error by adopting the same value (Eq. 2.11) of Σ as before, and by taking the same values (Eq. 2.9) of T_{CMB} and $\overline{T}$. Of course, prior to the birth of the inner core, $\mathcal{Q}^L = \mathcal{Q}^G = \mathcal{H}_{ICB} = 0$, and the dynamo is entirely thermally driven. Then Eqs (4.1) and (4.12) are replaced by

$$\mathcal{Q}_{CMB} = \mathcal{Q}^S + \mathcal{Q}^R, \tag{4.26}$$

$$\frac{\mathcal{Q}_{CMB}}{T_{CMB}} = \Sigma + \frac{\mathcal{Q}^D}{T_D} + \frac{\mathcal{Q}^S + \mathcal{Q}^R}{\overline{T}}. \tag{4.27}$$

The first of these is no more than a convenient way of defining the rate of cooling.

We may write Eqs (4.26) and (4.27) in the forms (4.24) and (4.25), with $\mathcal{H}_{ICB}$ set zero and different values of the constants E_λ, D_0 and D_λ. In our standard model, we have

$$\mathcal{Q}_{CMB} = \frac{2.31\,\text{TW}}{\lambda} + \mathcal{Q}^R, \tag{4.28}$$

$$\mathcal{Q}^D = \frac{T_D}{T_{CMB}}\left(\frac{0.28\,\text{TW}}{\lambda} - 0.68\,\text{TW} + 0.12\mathcal{Q}^R\right). \tag{4.29}$$

The parameter λ has, of course, a different interpretation than in Sections 4.1–4.3, where it referred to the growth rate of the inner core. In Eqs (4.28) and (4.29), 1.2λ is the timescale, $\tau_{cooling}$, over which the fluid core cools.

Taking $\mathcal{Q}^R = 0$ and $T_D = \overline{T}$ as before, we find that $\tau_{cooling} = 0.49\,\text{Gyr}$, and that the limits (2.19) give

$$12.9\,\text{TW} < \mathcal{Q}_{CMB} < 20.1\,\text{TW}, \quad 0.22\,\text{Gyr} > \tau_{cooling} > 0.14\,\text{Gyr}. \tag{4.30}$$

The other values of ξ_a explored in Table 3 give similar results. The small limits on $\tau_{cooling}$ given in Eq. (4.30) are not encouraging for the success of the dynamo when the inner core is absent; primordial heat must be lost quickly to drive a dynamo of the required strength and whether the mantle is able to transmit so much heat is uncertain.

5. Thermodynamic efficiency

There is no unique definition of the efficiency of the geodynamo, considered as a heat engine. Nevertheless, it is tempting to define the useful work to be the part $\mathcal{Q}^{\mathcal{J}}$ of $\mathcal{Q}^D$ that can be estimated through geodynamo simulations. Then the efficiency of the Dynamo is

$$\varepsilon_D \equiv \mathcal{Q}^{\mathcal{J}}/\mathcal{Q}_{CMB}, \tag{5.1}$$

which we write as

$$\varepsilon_D = f_F \varepsilon_G, \tag{5.2}$$

where ε_G is the ideal Geodynamo efficiency,

$$\varepsilon_G \equiv \mathcal{Q}^D/\mathcal{Q}_{CMB}, \tag{5.3}$$

and

$$f_F = \mathcal{Q}^{\mathcal{J}}/\mathcal{Q}^D \tag{5.4}$$

is a "frictional factor." By Eq. (4.19), we therefore have (taking $\mathcal{Q}^R = 0$ and $T_D = \overline{T}$ as before)

$$\varepsilon_G = \left(\frac{\overline{T}}{T_{CMB}} - 1\right) + \frac{1}{\mathcal{Q}_{CMB}}\left[-\Sigma\overline{T} + \mathcal{Q}^G + (\mathcal{H}_{ICB} + \mathcal{Q}^L)\left(1 - \frac{\overline{T}}{T_{ICB}}\right)\right]. \tag{5.5}$$

Our estimates of Section 4 give (for $\lambda = 1$)

$$\varepsilon_G \approx 17\%, \quad \varepsilon_D \approx 4\%, \quad f_F \approx 0.25. \tag{5.6}$$

A similar conclusion follows from an expression[2] derived by Braginsky and Roberts (1995):

$$\varepsilon_G \approx \frac{1}{\mathcal{Q}_{CMB}}\left[\mathcal{Q}^G + \left(\frac{\overline{T}}{T_{CMB}} - 1\right)(\mathcal{Q}_{sa} + \mathcal{H}_{ICB} + \mathcal{Q}^L)\right]; \tag{5.7}$$

this is about 18%. Equation (5.7) gives full force to the gravitational source but the two remaining (thermal) sources are each diminished in usefulness by the factor $(\overline{T}/T_{\text{CMB}} - 1) \approx 0.14$.

It was remarked in Section 4.1 that the parameter values of Braginsky and Roberts (1995) gave much the same $\mathcal{Q}_{\text{CMB}}$ as here, but that the individual contributions to $\mathcal{Q}_{\text{CMB}}$ were different. Here the thermal driving is larger and the dynamo is therefore less efficient; in the case of Braginsky and Roberts (1995), both Eqs (5.6) and (5.7) gave $\varepsilon_G \approx 28\%$ and $f_F \approx 0.15$.

When there is no inner core, Eq. (5.5) reduces (for $\mathcal{Q}^R = 0$ and $T_D = \overline{T}$) to

$$\varepsilon_G = \left(\frac{\overline{T}}{T_{\text{CMB}}} - 1 \right) - \frac{\Sigma\overline{T}}{\mathcal{Q}_{\text{CMB}}}. \tag{5.8}$$

For order of magnitude purposes, we take $\Sigma\overline{T}/\mathcal{Q}_{\text{CMB}} = 0.11$ as before, and obtain

$$\varepsilon_G \approx 3\%, \tag{5.9}$$

which confirms the difficulty faced when the inner core is absent and there is no radioactivity.

6. The radioactive core

6.1. Where is the potassium?

As the Sun contains 99.9% of the mass in the solar system, it is reasonable to suppose that its composition resembles that of the early planetary nebula. In confirmation, it has been observed that the CI chondrite meteoritic class, which represents the most primitive of all the undifferentiated meteorites, is unique in having a composition close to that of the Sun (Grevesse *et al.*, 1996). It has long been argued the Earth's composition is broadly chondritic (e.g. Ringwood, 1959, 1975, 1979; Wanke, 1981; Sun, 1982; Wasson and Kallemeyn, 1988), at least with respect to the refractory elements. Conventional wisdom holds that, because of the progressive decrease in the average mid-plane temperature of the nebula with increasing heliocentric distance, volatile elements in gaseous phase near the Sun would evaporate and be blown outwards by the solar wind, which was particularly violent during the early, T-Tauri stage of the Sun's evolution. This explains the change from the rocky inner planets (Mercury, Venus, Earth, Mars) to the giant gaseous outer planets (Jupiter, Saturn, Neptune, Uranus). For this reason, most modern workers would argue that the Earth's composition is that of a devolatilized CI chondrite (e.g. Hart and Zindler, 1986, 1989; McDonough and Sun, 1995).

This idea is consistent with estimates of the concentrations of uranium, thorium and potassium in the bulk silicate Earth (BSE), the name given to the homogeneous reservoir representing the sum of the continental crust, continental lithosphere and the depleted mantle reservoirs, which taken together and appropriately weighted give

$$U_{\text{BSE}} = 11.92\,\text{ppb}, \quad Th_{\text{BSE}} = 46.13\,\text{ppb}, \quad K_{\text{BSE}} = 140\,\text{ppm} \tag{6.1}$$

(ppb = parts per billion; ppm = parts per million).

It is generally agreed that U and Th cannot exist in the core and, if the same is true for K, we may translate Eq. (6.1) into concentrations for the bulk Earth (BE), that is, the Earth as a whole, by multiplying the entries by $\omega = 0.670$, the ratio of the mass of the BSE to that of the entire Earth:

$$U_{\text{BE}} = 8.0\,\text{ppb}, \quad Th_{\text{BE}} = 30.9\,\text{ppb}, \quad K_{\text{BE}} = 93.7\,\text{ppm}. \tag{6.2}$$

These figures may be compared with the average CI chondrite values (e.g. McDonough and Sun, 1995):

$$U_{CI} = 8.1\,\text{ppb}, \quad Th_{CI} = 28.9\,\text{ppb}, \quad K_{CI} = 555\,\text{ppm}. \tag{6.3}$$

The U/Th ratios are very similar: 0.26 for BE and 0.28 for CI, which are well within the uncertainties of estimation, but the K/U ratios are very different: 1.2×10^4 for a potassium depleted BE but 6.7×10^4 for CI.

There are two extreme explanations for this fact, and any number of intermediate possibilities. One extreme is the traditional view, which is today widely held by the geochemical and cosmochemical community: potassium is a moderately volatile element and the missing K was lost during the formation of the Earth. The other extreme is the unorthodox view that all the missing K resides in the core.

The traditional view depends on the belief that equilibrium liquid silicate/liquid metal partitioning forbids any K from entering the core. This view is currently so strongly held that it is used as a way of placing a lower bound of 1,100 K on the temperature of the solar nebula at the time of Earth's formation, this being the 50% condensation temperature of K (Larimar, 1988; Palme *et al.*, 1988). The unorthodox view, which at one time was the traditional view (e.g. Murthy and Hall, 1970, 1972; Lewis, 1971; Goettel, 1972, 1974, 1976), was abandoned because it was argued, mostly on the basis of low pressure partitioning experiments (e.g. Goettel, 1972; Somerville and Ahrens, 1980; Ito *et al.*, 1993; Chabot and Drake, 1999) that K cannot form a stable alloy with Fe. Without contesting this, Calderwood (2000) has pointed out that potassium can enter the core by another channel, as explained in the next subsection.

The traditional view rules out the presence of ^{40}K as a heat source for the dynamo, leading to the difficulties outlined in Section 5.

6.2. The magma ocean: the metallization of potassium

Planetary accretion modelers visualize that, after the proto-Earth reaches 1/3–1/2 of its present mass, heating by the impact of planetesimals will create a hydrous global magma ocean (HGMO); see Davies (1985, 1990). Subsequently, as the Earth cooled, mantle–core segregation took place at the base of the HGMO under the conditions of very high temperatures and pressures prevailing there (e.g. Thibault and Walter, 1995; Li and Agee, 1996, 1997; Righter and Drake, 1996, 1997, 1999; Suzuki *et al.*, 1996; Righter *et al.*, 1997, 2000; Wade and Wood, 2001). In this model, liquid iron undergoes gravitational separation from the liquid silicate magma and ponds at the base of the magma ocean. Then, via percolation through the underlying solid mantle (e.g. Shannon and Agee, 1998), this liquid metal continues to migrate deeper into the Earth. This model is invoked to explain the observed mantle depletions in Co, Ni, W, P, V, S, Pb, Ga and Nb within the BSE relative to CI chondritic values, the argument being that the present day mantle abundancies do not represent true depletions but instead record equilibrium element partitioning during mantle–core segregation (see above references). To obtain present day BSE concentrations of these elements under equilibrium partitioning, the prevailing pressures and temperatures would have to be about

$$P_{HGMO} = 22\,\text{GPa}, \quad T_{HGMO} = 2,200\,\text{K}, \tag{6.4}$$

the former corresponding roughly to the bottom of the present Earth's transition zone; see Righter and Drake (1999).

The unorthodox view of Calderwood (2000) rests on the experimental partitioning coefficients of Ohtani *et al.* (1997). Their experiments included nickel in the metal fraction, and went to a high enough pressure (20 GPa) to exceed comfortably the pressure (P >11.5 GPa) at which K has been experimentally observed to undergo an s to d electronic transition, termed "metallization" (e.g. Winzenick *et al.*, 1994). This change in the electronic configuration of K is critical because it has been shown in separate diamond anvil experiments that, once K has undergone metallization, it will form a stable metal alloy with Ni (Parker *et al.*, 1996); but at pressures beneath this pressure interval, no stable alloy forms. Parker *et al.* (1996) point out that their results are supported by empirical chemical rules, termed Miedema rules (Miedema *et al.*, 1980), that explain the formation of transition metal alloys and the likelihood of alloy formation with increasing pressure. This is in sharp contrast to a possible potassium–iron metal alloy, which is neither observed at low to high pressures (e.g. Ito *et al.*, 1993; Chabot and Drake, 1999), nor is predicted to form according to the Miedema Rules (Parker *et al.*, 1996; see also Bukowinski, 1976; Liu, 1986; Sherman, 1990). In short, both theory and diamond anvil experiments show that potassium will readily form a high pressure alloy with nickel, but not with iron.

With the exception of Ohtani *et al.* (1997), previous high partitioning experiments designed to test if potassium might partition into the core did not include nickel in the starting metal phase and instead concentrated solely on either pure iron or FeS mixtures for the metal phase. Thus, it is not surprising that these experiments found very little partitioning of K into the metal phase.

The unorthodox view evades a difficulty encountered by the traditional view. If (see above) the solar nebula at 1 au had been at 1,100 K, not only should the K have been depleted but also the $^{41}K/^{39}K$ isotopic ratio should be been increased relative to the incoming planetesimals (assumed of chondritic composition), and such an effect has never been observed (Humayun and Clayton, 1995a,b). Furthermore, a temperature of 1,100 K would predict an Earth even more depleted in elements that are more volatile than potassium, such as Pb, for which the 50% condensation temperature is 525 K and S for which it is 650 K; see Boss (1998).

6.3. Some consequences of potassium in the core

We pursue the unorthodox view. If we put all the missing K into the core, its concentration would, according to Eqs (6.2) and (6.3), be $(555 - 93.7\omega)/(1 - \omega) \approx 1,420$ ppm. The $1.95 \times 10^{24} \times 1.420 \times 10^{-3} \approx 2.82 \times 10^{21}$ kg of K in the core would, through the ^{40}K it contains, generate 3.391×10^{-9} W kg^{-1} of heat, or about 9.34 TW in total. The half-life of ^{40}K is 1.2778 Gyr, so that its heat production 4.5 billion years ago would be twelve times as great. We at first ignore this fact.

A source of radioactivity in the core can slow down, halt, or even reverse the growth of the inner core; this depends entirely on Q_{CMB}, the heat flux from the core into the mantle, and this in turn depends on the effectiveness of mantle convection in transporting heat. The mantle is the "valve" that controls magnetoconvection in the core and the geodynamo.

When we allow for $Q^R = 9.34$ TW of radiogenic heat in Eqs (4.10) and (4.21), we find that there is no longer a critical λ for which Q^D is zero; now Q^D is always positive. The inequality (2.19) gives

$$11.1 \text{ TW} < Q_{CMB} < 14.7 \text{ TW}, \quad 5.3 \text{ Gyr} > \tau_{SIC} > 1.6 \text{ Gyr}. \tag{6.5}$$

This range of τ_{SIC} includes the age of the Earth, that is, it is possible that the SIC has existed over most of geological time and has produced a dynamo for which 1 TW $< Q^J < 2$ TW. This is fortunate, because the thermodynamic obstacles to dynamo action in a totally fluid core (Section 4.4) are not much eased by the radioactive sources. For such a core, the bounds (2.19) give

$$12.9 \, \text{TW} < \mathcal{Q}_{\text{CMB}} < 20.1 \, \text{TW}, \quad 0.76 \, \text{Gyr} > \tau_{\text{cooling}} > 0.25 \, \text{Gyr}, \quad (6.6)$$

which may be compared with range (4.30). Moreover the core must rid itself of heat very rapidly, and the mantle must transmit it. The same is even true, to a lesser extent, when the inner core is present and range (6.5) holds.

Let us suppose that the SIC has existed for 4 Gyr ($\lambda = 10/3$), and let us continue to suppose that $\mathcal{Q}^R$ is unchanging. We find that

$$\mathcal{Q}_{\text{CMB}} = 11.6 \, \text{TW}, \quad \mathcal{Q}^D = 1.1 \, \text{TW}. \quad (6.7)$$

(If $\mathcal{Q}^R$ is halved, we find that, for $\tau_{\text{SIC}} = 4 \, \text{Gyr}$, $\mathcal{Q}_{\text{CMB}} = 7.2 \, \text{TW}$, as assumed by Glatzmaier and Roberts in their simulations; then $\mathcal{Q}^D = 0.53 \, \text{TW}$.) In terms of the efficiencies introduced in Section 5, Eq. (6.8) gives

$$\varepsilon_G \approx 9\%, \quad \varepsilon_D \approx 0.3\%, \quad f_F \approx 0.27. \quad (6.8)$$

The reduction of efficiency, as compared with Eq. (5.6) reflects the effect of the added thermal forcing created by $\mathcal{Q}^R$.

If we allow for the greater strength of the radioactive sources 4 Gyr ago, we find that $\mathcal{Q}_{\text{CMB}} \approx \mathcal{Q}^R \approx 82 \, \text{TW}$ and $\mathcal{Q}^D \approx 11 \, \text{TW}$, which, if the mantle is able to transmit so much heat, leaves abundant energy to drive a very strong dynamo. For such large values of $\mathcal{Q}^R$ both the pair of equations (4.24) and (4.25) and the pair (4.28) and (4.29) reduce approximately to

$$\mathcal{Q}_{\text{CMB}} = \mathcal{Q}^R, \quad \mathcal{Q}^D = 0.12 \frac{T_D}{T_{\text{CMB}}} \mathcal{Q}^R \quad (6.9)$$

and the presence or absence of an inner core has no thermodynamic significance.

7. Conclusions

At first sight, it appears that the ohmic dissipation associated with the geomagnetic field is rather small in comparison with the thermal and gravitational energy sources available to drive the dynamo. As we saw in the Section 2.2, less than 6 MW of dissipation is the minimum sufficient to sustain the present dipole field of the Earth, but there is no less than 44 TW of heat energy coming out of the Earth's surface. There is surely a huge "comfort-zone" here removing any worries that there is insufficient energy to drive the dynamo. Unfortunately, there are a large number of factors that chip away at this comfort-zone. First, the 6 MW assumes the field is in an optimal configuration to reduce ohmic dissipation. A more realistic configuration, such as the interior field having the form of a free decay mode, increases the dissipation to around 80 MW. Recent simulations of the geodynamo suggest that the invisible component of the field which never leaves the core (consisting of both toroidal and poloidal components) is much larger than the 0.0005 T field escaping at the CMB. A toroidal field of twenty times this strength, 0.01 T, is typical of what is found. This increases the dissipation by a factor of 400, and an additional factor of 2 is provided by the slightly higher dissipation rate of toroidal modes compared to poloidal modes. Thus the dissipation from the largest scale modes could well be around 0.08 TW.

The next factor to take into account is the contribution to ohmic dissipation from the smaller scale modes. As we argue in Section 2, the modes in the spherical harmonic range $n \sim 10\text{--}40$ are the most likely to be providing the bulk of the dissipation, although this depends on the form of the magnetic power spectrum. Not much is known about this at present, though it could be a fruitful field for further investigation. Our present estimates are based on the simulations of Roberts and Glatzmaier (2000a), and these suggest that the 0.08 TW large-scale dissipation could be enhanced up to 1 or 2 TW by the smaller-scale components. Thus, we see that the 6 MW "minimum" dissipation is a gross underestimate of the actual dissipation required to produce a 0.0005 T field coming out of the CMB. If viewed as a means of producing external dipole field, the dynamo process is extraordinarily inefficient, with most of the generated field never emerging at all, and with small scale "useless" components dissipating most of the magnetic energy.

There are still nevertheless 44 TW of heat coming out of the Earth; surely this is sufficient to produce 1–2 TW of ohmic dissipation? There are two main problems here. First, on the traditional view, only about 7 TW (or may be less) of this 44 TW pass through the CMB. The rest is produced by radioactivity in the mantle, useless for the dynamo. The thermal heat flux is also subject to the "Carnot efficiency factor." Since the temperature difference between the CMB and the ICB is only around 20% of the actual temperature, only a small fraction of the available heat flux can go into ohmic dissipation. The gravitational energy associated with compositional convection is not subject to this efficiency factor tax, but unfortunately the amount of energy available from this source now appears to be less than originally thought, owing to new work on the crystallization process (Section 3.4). The net effect of these considerations is that it now appears that the energy budget for the geodynamo is rather tight, and we must start to consider ways in which either the process could be more efficient than we currently believe, or that there are new energy sources such as radioactivity in the core.

The energy budget calculations depend on a wide range of geophysical parameters, some of which are well known, while others remain uncertain. We have revised the estimates given in Braginsky and Roberts (1995) in a number of ways. Their estimates were made using the PREM model, while here we have used the ak135 model of Kennett *et al.* (1995). The resulting differences are rather small, and do not affect the estimates of the various components of the energy and entropy fluxes very greatly. The heat flux conducted down the adiabat is significant in the core, so the energetics are sensitive to the values of the thermal conductivity assumed. These values are constantly revised as our knowledge of high-pressure material physics improves, but the changes over the last 6 years have not led to very significant changes to the gross energy balance. The most significant uncertainties are connected with the freezing process and the growth of the inner core. The jump in density between the SIC and the FOC at the ICB boundary has been fairly well established by seismology, but this jump consists of two parts, one due to the contraction of iron on solidifying and one due to the release of light material. It now appears that the part due to contraction on solidification was previously underestimated, while correspondingly, the release of light material was overestimated. In consequence, the latent heat of crystallization has been revised upward (Anderson and Duba, 1997) leading to a significant increase in the estimate of total rate of latent heat release at the ICB, but the rate of gravitational energy release is now less than previously thought. Although the overall energy release is much the same as in the previous estimates, this change reduces the energy available for ohmic dissipation, because the Carnot efficiency factor bites into the extra thermal energy released by latent heat, but we lose the full amount of the decrease in gravitational energy. In consequence, if the slower estimates of core cooling are adopted, corresponding to the inner core being in existence for 3 Gyr, there is a danger that there is insufficient energy to drive a dynamo at all.

There is a particular problem with the early Earth. Paleomagnetic data reaching back 3 Gyr is of course far less certain than that accumulated for more recent epochs, but if the work of Kono and Tanaka (1995) is confirmed, dynamo theory will have to take on board the possibility that a geomagnetic field not dissimilar to the present field can be generated with no inner core. From the discussion in Section 4.3 we see that this makes the energy problem much more acute. To obtain a vigorous 1–2 TW dynamo with no inner core requires a great deal more heat flux through the CMB than the 7 TW the current standard model suggests.

There are a number of ways in which the energy problem could be alleviated. Our knowledge of the composition of core material and of the high-pressure physics that governs the behavior near the ICB is still very incomplete, and it is possible that new investigations may increase estimates of gravitational release, alleviating the problem at least when there is a growing inner core. We should also remember that the field of dynamo simulation is still in its infancy, and it may be that more efficient dynamos emerge, allowing us to reduce the amount of ohmic dissipation while still maintaining an external field of the observed strength. This issue has not received very much attention to date, and further progress may well be possible. Whether new developments will improve the situation (or make it worse!) remains to be seen.

There is, however, another way of overcoming the difficulty if radioactivity is allowed in the core. There is then no problem in finding the energy required to drive a dynamo with or without an inner core. This may also have implications for other planetary dynamos, such as the recently discovered dynamo in Ganymede. In Section 6 we have presented the case for radioactive potassium entering the core via a nickel alloy. At present, our understanding of the planetary formation process is still very rudimentary; no doubt this important issue of the amount of radioactivity in planetary cores, which has profound implications for the thermal history of the solar system, will be further discussed by the geochemistry community. Another possible way of deciding this issue would be through mantle convection studies. The heat coming out of the core must pass through the thermal boundary layer at the base of the mantle, and then be convected up through the mantle. Observations of the thermal boundary layer, which is believed to be the D″ layer, or mantle convection simulations may be able to place useful constraints on the amount of heat going through the CMB.

Acknowledgments

One of us (P.H.R.) was supported by the National Science Foundation with grant NSF EAR97-25627. A.C. was supported by a DOE Cooperative Agreement (#DE-FCO8-98NU13410) between the High Pressure Science Center at the University of Nevada-Las Vegas, Department of Physics, and the US Department of Energy.

Notes

1 Loper's method of computing Eq. (4.6) leads to a value of Q^S almost twice as large as ours. The difference must be attributed to contraction of the CMB, which is included in his model but not ours.
2 Their result followed from the energy and entropy balances for the whole core using their modified Boussinesq theory. It also depended on a coincidence which makes it inappropriate here for Section 6: in their model, $2(1 - T_{CMB}/\overline{T}) \approx 1 - T_{CMB}/T_{ICB}$. We use balances for the FOC alone, and have therefore again allowed for the fluxes across the ICB by replacing Q^L by $Q^L + Q_{CMB} \approx Q^L + \mathcal{H}_{CMB}$. To obtain Eq. (5.7) from their expression, we write $T_D = \overline{T}$, as before.

References

Abramowitz, M. and Stegun, I. A. Eds, *Handbook of Mathematical Functions with Formulas, Graphs and Mathematical Tables*, NBS Appl. Math. Ser., #55 (1964).

Anderson, O. L. and Duba, A., "Experimental melting curve of iron revisited," *J. Geophys. Res.* **102**, 22659–22669 (1997).

Backus, G. E., "Gross thermodynamics of heat engines in the deep interior of the Earth," *Proc. Nat. Acad. Sci. Wash.* **72**, 1555–1558 (1975).

Backus, G., Parker, R. and Constable, C. G., *Foundations of Geomagnetism*. Cambridge University Press, Cambridge (1996).

Boehler, R., "Experimental constraints on melting conditions relevant to core formation," *Geochim. Cosmochim. Acta* **60**, 1109–1112 (1996).

Boss, A. P., "Temperatures in protoplanetary disks," *Annu. Rev. Earth Planet. Sci.* **26**, 53–80 (1998).

Braginsky, S. I., "Magnetohydrodynamics of the Earth's core," *Geomag. Aeron.* **4**, 898–916 (1964).

Braginsky, S. I. and Meytlis, V. P., "Local turbulence in the Earth's core," *Geophys. Astrophys. Fluid Dynam.* **55**, 71–87 (1990).

Braginsky, S. I. and Roberts, P. H., "Equations governing Earth's core and the geodynamo," *Geophys. Astrophys. Fluid Dynam.* **79**, 1–97 (1995).

Buffett, B. A., Huppert, H. E., Lister, J. R. and Woods, A. W., "On the thermal evolution of the Earth's core," *J. Geophys. Res.* **101**, 7989–8006 (1996).

Bukowinski, M. S. T., "The effect of pressure on the physics and chemistry of potassium," *Geophys. Res. Lett.* **3**, 491–503 (1976).

Calderwood, A. R., "The distribution of U, Th, and K in the Earth's crust, lithosphere, mantle and core: constraints from an elemental mass balance model and the present day heat flux," In: *SEDI 2000*, the 7th Symposium on Studies of the Earth's Deep Interior, Exeter, UK, Abstract Vol. S9.4 (2000).

Chabot, N. L. and Drake, M. J., "Potassium solubility in metal: the effects of composition at 15 kilobars and 1900°C on partitioning between iron alloys and silicate melts," *Earth Planet. Sci. Lett.* **172**, 323–335 (1999).

Davies, G. F., "Heat deposition and retention in a solid planet growing by impacts," *Icarus* **63**, 45–68 (1985).

Davies, G. F., "Heat and mass transport in the early Earth," In: *Origin of the Earth* (Eds J. H. Jones, and H. E. Newsom), pp. 175–194. Oxford University Press, New York (1990).

Dziewonski, A. M. and Anderson, D. L., "Preliminary reference Earth model," *Phys. Earth Planet. Inter.* **25**, 297–356 (1981).

Glatzmaier, G. A. and Roberts, P. H., "A three-dimensional convective dynamo solution with rotating and finitely conducting inner core and mantle," *Phys. Earth Planet. Inter.* **91**, 63–75 (1995a).

Glatzmaier, G. A. and Roberts, P. H., "A three-dimensional self-consistent computer simulation of a geomagnetic field reversal," *Nature* **377**, 203–209 (1995b).

Glatzmaier, G. A. and Roberts, P. H., "An anelastic evolutionary geodynamo simulation driven by compositional and thermal convection," *Physica D* **97**, 81–94 (1996a).

Glatzmaier, G. A. and Roberts, P. H., "Rotation and magnetism of Earth's inner core," *Science* **274**, 1887–1891 (1996b).

Glatzmaier, G. A. and Roberts, P. H., "Simulating the geodynamo," *Contemp. Phys.* **38**, 269–288 (1997).

Glatzmaier, G. A., Coe, R. S., Hongre, L. and Roberts, P. H., "The role of the Earth's mantle in controlling the frequency of geomagnetic reversals," *Nature* **401**, 885–890 (1999).

Goettel, K. A., "Partitioning of potassium between silicates and sulphide melts: experiments relevant to the Earth's core," *Phys. Earth Planet. Inter.* **6**, 161–166 (1972).

Goettel, K. A., "Potassium in the Earth's core: evidence and implications," In: *Physics and Chemistry of Minerals and Rocks* (Ed. R.G. Strews Jr), pp. 479–489. John Wiley Interscience, New York (1974).

Goettel, K. A., "Models for the origin and composition of the Earth, and the hypothesis of potassium in the Earth's core," *Geophys. Surv.* **2**, 36–397 (1976).

Grevesse, N., Noels, A. and Sauval, A. J., "Standard abundances," In: *Cosmic Abundances* (Eds S. S. Holt and G. Sonneborn), *Astronom. Soc. of the Pacific Conf. Ser.* **99**, 117–126 (1996).

Gubbins, D., "Energetics of the Earth's core," *J. Geophys.* **43**, 453–464 (1977).

Gubbins, D., Masters, T. G. and Jacobs, J. A., "Thermal evolution of the Earth's core," *Geophys. J. R. Astr. Soc.* **59**, 57–99 (1979).

Hart, S. R. and Zindler, A., "In search of a bulk Earth composition," *Chem. Geol.* **57**, 247–267 (1986).

Hart, S. R. and Zindler, A., "Constraints on the nature and development of chemical heterogeneities in the mantle," In: *Mantle Convection* (Ed. W. R. Peltier), pp. 261–387. Gordon and Breach Science Publishers, New York (1989).

Hewitt, J. M., McKenzie, D. P. and Weiss, N. O., "Dissipative heating in convective flows," *J. Fluid Mech.* **68**, 721–738 (1975).

Humayun, M. and Clayton, R. N., "Precise determination of the isotopic composition of potassium: Application to terrestrial rocks and lunar soils," *Geochim. Cosmochim. Acta*, **59**, 2115–2130 (1995a).

Humayun, M. and Clayton, R. N., "Potassium isotope cosmochemistry: genetic implications of volatile element depletion," *Geochim. Cosmochim. Acta* **59**, 2131–2148 (1995b).

Ito, E., Morooka, K. and Ujike, O., "Dissolution of K in molten iron at high pressure and temperature," *Geophys. Res. Lett.* **20**, 1651–1654 (1993).

Jeanloz, R., "The nature of the Earth's core," *Annu. Rev. Earth Planet. Sci.* **18**, 357–386 (1990).

Jones, C. A. and Roberts, P. H., "Convection driven dynamos in a rotating plane layer," *J. Fluid Mech.* **404**, 311–343 (2000).

Kennett, B. L. N., Engdahl, E. R. and Buland, R., "Constraints on seismic velocities in the Earth from traveltimes," *Geophys. J. Int.* **122**, 108–124 (1995).

Kono, M. and Tanaka, H., "Intensity of the geomagnetic field in geological time: a statistical study," In: *The Earth's Central Part: Its Structure and Dynamics* (Ed. T. Yukutake), pp. 75–94. Terrapub, Tokyo, Japan (1995).

Labrosse, S., Poirier, J.-P. and Mouël, J.-L., "On the cooling of Earth's core," *Phys. Earth Planet. Inter.* **99**, 1–17 (1997).

Labrosse, S., Poirier, J.-P. and Mouël, J.-L., "The age of the inner core," *Earth Planet Sci. Letts.* **5895**, 1–13 (2001).

Laio, A., Bernard, S., Chiarotti, G. L., Scandolo S. and Tosatti, E., "Physics of iron at Earth's core conditions," *Science* **287**, 1027–1030 (2000).

Langel, R. A., "The main field," In: *Geomagnetism* **1**, (Ed. J. A. Jacobs). Academic Press, New York (1987).

Larimer, J. W., "The cosmochemical classification of the elements," In: *Meteorites and the Early Solar System* (Eds J. F. Kerridge and M. S. Mathews), pp. 436–461. University of Arizona Press, Tucson, AZ (1988).

Lewis, J. S., "Consequences of the presence of sulfur in the core of the Earth," *Earth Planet. Sci. Letts.* **11**, 130–134 (1971).

Li, J., and Agee, C. B., "Geochemistry of mantle-core differentiation at high pressure," *Nature* **381**, 687–689 (1996).

Li, J., and Agee, C. B., "Partitioning of volatile elements during core formation," In: *Seventh Annual V. M. Goldschmidt Conference*, LPI Contribution No. 921, p. 126. Lunar and Planetary Institute, Houston TX (1997).

Liu, L.-G., "Potassium and the Earth's core," *Geophys. Res. Lett.* **13**, 1145–1148 (1986).

Loper, D. E., "The gravitationally powered dynamo," *Geophys. J. R. Astr. Soc.* **54**, 389–404 (1978).

Loper, D. E., "The nature and consequences of thermal interactions twixt core and mantle," *J. Geomag. Geoelectr.* **43**, 79–91 (1991).

McDonough, W. F. and Sun, S.-S., "The composition of the Earth," *Chem. Geol.* **120**, 223–253 (1995).

Merkel, S., Goncharov, A. F., Mao, H.-K., Gillet, Ph. and Hemley, R. J., "Raman spectroscopy of iron to 152 GigaPascals: Implications for Earth's inner core," *Science* **288**, 1626–1629 (2000).

Miedema, A. R., de Chatel, P. F. and de Boer, F. R., "Cohesion in alloys – fundamentals of a semi-empirical model," *Physica* **B100**, 1–28 (1980).

Moffatt, H. K., *Magnetic Field Generation in Electrically Conducting Fluids*. Cambridge University Press, Cambridge (1978).

Murthy, V. R. and Hall, H. T., "The chemical composition of the Earth's core: possibility of sulfur in the core," *Phys. Earth Planet. Inter.* **2**, 276–282 (1970).

Murthy, V. R. and Hall, H. T., "The origin and chemical composition of the Earth's core," *Phys. Earth Planet. Inter.* **6**, 123–130 (1972).

Ohtani, E., Yurimoto, H. and Seto, S., "Element partitioning between metallic liquid, silicate liquid, and lower mantle minerals: implications for core formation of the Earth," *Phys. Earth Planet. Inter.* **100**, 97–114 (1997).

Palme, H., Larimer, J. W. and Lipshutz, M. E., "Moderately volatile elements," In: *Meteorites and the Early Solar System* (Eds J. F. Kerridge, and M. S. Mathews), pp. 436–461. University of Arizona Press, Tucson, AZ (1988).

Parker, L. J., Atou, T. and Badding, J. V., "Transition element-like chemistry for potassium under pressure," *Science* **273**, 95–97 (1996).

Poirier, J.-P. and Shankland, T. J., "Dislocation melting of iron and the temperature of the inner core," *Geophys. J. Int.* **115**, 147–151 (1993).

Righter, K. and Drake, M. J., "Core formation in Earth's Moon, Mars, and Vesta," *Icarus* **124**, 513–529 (1996).

Righter, K. and Drake, M. J., "Metal-silicate equilibrium in a homogeneously accreting Earth: new results for Re," *Earth Planet. Sci. Lett.* **146**, 541–553 (1997).

Righter, K. and Drake, M. J., "Effect of water on metal–silicate partitioning of siderophile elements: a high pressure and temperature terrestrial magma ocean and core formation," *Earth Planet. Sci. Lett.* **171**, 383–399 (1999).

Righter, K. and Drake, M. J. and Yaxley, G., "Prediction of siderophile element metal-silicate partition coefficients to 20 GPa and 2800°C: the effects of pressure, temperature, oxygen fugacity, and silicate and metallic melt compositions," *Phys. Earth Planet.* **100**, 115–134 (1997).

Righter, K., Walker R. J. and Warren, P. H., "Significance of highly siderophile elements and osmium isotopes in the lunar and terrestrial mantles," In: *Origin of the Earth and Moon* (Eds R. M. Canup and K. Righter), pp. 291–322. University of Arizona Press, Tucson, AZ (2000).

Ringwood, A. E., "On the chemical evolution and densities of the planets," *Geochim. Cosmochim. Acta* **51**, 257–287 (1959).

Ringwood, A. E., *Composition and Petrology of the Earth's Mantle*. McGraw Hill, New York, (1975).

Ringwood, A. E., *Origin of the Earth and Moon*. Springer, New York (1979).

Roberts, P. H. and Glatzmaier, G. A., "A test of the frozen flux approximation using geodynamo simulations," *Phil. Trans. R. Soc. Lond.* A **358**, 1109–1121 (2000a).

Roberts, P. H. and Glatzmaier, G. A., "Geodynamo theory and simulations," *Rev. Mod. Phys.* **72**, 1081–1123 (2000b).

Sarson, G. R. and Jones, C. A., "A convection driven geodynamo reversal model," *Phys. Earth Planet. Inter.* **111**, 3–20 (1999).

Shannon, M. C. and Agee, C. B., "Percolation of core melts at lower mantle conditions," *Science* **280**, 1059–1061 (1998).

Shearer, P. and Masters, G., "The density and shear velocity contrast at the inner core boundary," *Geophys. J. Int.* **102**, 491–496 (1991).

Sherman, D. M., "Chemical bonding and the incorporation of potassium into the Earth's core," *Geophy. Res. Lett.* **17**, 693–696 (1990).

Somerville, M. and Ahrens, T. J., "Shock compression of $KFeS_2$ and the question of potassium in the core," *J. Geophy. Res.* **85**, 7016–7024 (1980).

St Pierre, M. G., "On the local nature of turbulence in Earth's outer core," *Geophys. Astrophys. Fluid Dynam.* **83**, 293–306 (1996).

Stacey, F. D., *Physics of the Earth*. Brookfield Press, Brisbane, 3rd edition (1992).

Stacey F. D. and Anderson, O. L., "Electrical and thermal conductivities of Fe–Ni–Si alloy under core conditions," *Phys. Earth Planet. Inter.* **124**, 153–162 (2001).

Stevenson, D. J., Spohn, T. and Schubert, G., "Magnetism and thermal evolution of the terrestrial planets," *Icarus* **54**, 466–489 (1983).

Sun, S.-S., "Chemical composition and origin of the Earth's primitive mantle," *Geochim. Cosmochim. Acta* **46**, 179–192 (1982).

Suzuki, T., Akaogi, M. and Yagi, T., "Pressure dependence of Ni, Co and Mn partitioning between iron hydride and olivine, magnesiownstite and pyroxene," *Phys Earth Planet. Inter.* **96**, 209–220 (1996).

Thibault, Y. and Walter, M. J., "The influence of pressure and temperature on the metal-silicate partition coefficients of nickel and cobalt in a model C1 chondrite and implications for metal segregation in a deep magma ocean," *Geochim. Cosmochim. Acta* **59**, 991–1002 (1995).

Wade, J. and Wood, B. J., "The Earth's 'missing' niobium may be in the core," *Nature* **409**, 75–78 (2001).

Wanke, H., "Constitution of terrestrial planets," *Phil. Trans. R. Soc. Lond.* A **303**, 287–302 (1981).

Wasson, J. T. and Kallemeyn, G. W., "Compositions of chondrites," *Phil. Trans. R. Soc. Lond.* A **325**, 535–544 (1988).

Williams, Q., Jeanloz, R., Bass, J., Svendsen, B. and Ahrens, T. J., "A melting curve for iron to 250 GigaPascals: a constraint on the temperature at Earth's center," *Science* **236**, 181–182 (1987).

Winzenick, M., Vijayakumar, V. and Holzapfel, W. B., "High pressure X-ray diffraction on potassium and rubidium up to 50 GPa," *Phys. Rev.* B **50**, 12381–12385 (1994).

6 Convection in rotating spherical shells and its dynamo action

Friedrich H. Busse, Eike Grote and Radostin Simitev

Institute of Physics, University of Bayreuth, D-95440 Bayreuth, Germany

1. Introduction

The evolution of celestial bodies such as the Earth is characterized by the transport of heat from the interior to outside. Typically the basic static (or nearly static) state in which heat is transported by conduction and radiation is unstable in all or in parts of the interior and convection flows occur. Unlike molecular conduction and radiation, convection flows are rather sensitive to the state of rotation of the body, unless the viscosity is very high as in the mantles of the terrestrial planets. The action of the Coriolis force on fluid motion usually inhibits the efficiency of the convective heat transport, and the ways in which oscillatory motions and turbulence may overcome the inhibiting influence of rotation pose some most interesting dynamical problems. A way chosen most frequently by nature to counteract the effects of strong rotation is the generation of a magnetic field. Through the Lorentz force a new participant enters the balance of forces and evidently facilitates a more efficient transport of energy.

In this chapter, we intend to discuss first the dynamical problems of convection and then turn to the roles that magnetic fields generated through the dynamo process may play in changing the structure of convection flows and their capacity for transporting heat. In Section 2 the basic equations and the numerical approach are introduced. In Section 3 we review briefly the onset of convection in rotating spherical fluid shells. The typical bifurcation scenarios that develop as the Rayleigh number increases and give rise to turbulent convection with its coherent structures are discussed in Section 4. In Section 5 the dynamo process in the presently computationally accessible parameter regime is discussed. Numerical solutions based on hyperdiffusivity schemes will not be considered since they tend to introduce artificial effects (Zhang and Jones, 1997; Grote *et al.*, 2000a). The interaction between magnetic fields and convection flows in the presence of a dominant Coriolis force is considered in Section 6. Open problems of future research are mentioned in Section 8.

2. Mathematical formulation of the problem

For the description of finite amplitude convection in rotating spherical shells and its dynamo action, we follow the standard formulation used in earlier work by the authors (Busse *et al.*, 1998; Grote *et al.*, 1999, 2000b). But we assume that a more general static state exists with the temperature distribution $T_S = T_0 - (\beta d^2 r^2/2) + \Delta T \eta r^{-1}(1-\eta)^{-2}$ where η denotes the ratio of inner to outer radius of the shell and d is its thickness. ΔT is the temperature difference between the boundaries in the special case $\beta = 0$. The gravity field is given by $\mathbf{g} = -\gamma d\mathbf{r}$ where $\mathbf{r}$ is the position vector with respect to the center of the sphere and r is its length measured in units of d. In addition to the length d, the time d^2/ν, the temperature $\nu^2/\gamma \alpha d^4$

and the magnetic flux density $v(\mu\varrho)^{1/2}/d$ are used as scales for the dimensionless description of the problem where v denotes the kinematic viscosity of the fluid, κ its thermal diffusivity, ϱ its density and μ is the magnetic permeability. The density is assumed to be constant except in the gravity term where its temperature dependence given by $\alpha \equiv (d\varrho/dT)/\varrho =$ constant is taken into account. Since the velocity field $\mathbf{u}$ as well as the magnetic flux density $\mathbf{B}$ are solenoidal vector fields, the general representation in terms of poloidal and toroidal components can be used

$$\mathbf{u} = \nabla \times (\nabla v \times \mathbf{r}) + \nabla w \times \mathbf{r}, \tag{1a}$$

$$\mathbf{B} = \nabla \times (\nabla h \times \mathbf{r}) + \nabla g \times \mathbf{r}. \tag{1b}$$

By multiplying the (curl)^2 and the curl of the Navier–Stokes equations in the rotating system by $\mathbf{r}$ we obtain two equations for v and w:

$$[(\nabla^2 - \partial_t)L_2 + \tau\partial_\varphi]\nabla^2 v + \tau Qw - L_2\Theta = -\mathbf{r} \cdot \nabla \times [\nabla \times (\mathbf{u} \cdot \nabla\mathbf{u} - \mathbf{B} \cdot \nabla\mathbf{B})], \tag{2a}$$

$$[(\nabla^2 - \partial_t)L_2 + \tau\partial_\varphi]w - \tau Qv = \mathbf{r} \cdot \nabla \times (\mathbf{u} \cdot \nabla\mathbf{u} - \mathbf{B} \cdot \nabla\mathbf{B}), \tag{2b}$$

where ∂_t and ∂_φ denote the partial derivatives with respect to time t and with respect to the angle φ of a spherical system of coordinates r, θ, φ and where the operators L_2 and Q are defined by

$$L_2 \equiv -r^2\nabla^2 + \partial_r(r^2\partial_r),$$

$$Q \equiv r \cos\theta \nabla^2 - (L_2 + r\partial_r)(\cos\theta\partial_r - r^{-1}\sin\theta\partial_\theta).$$

The heat equation for the dimensionless deviation Θ from the static temperature distribution can be written in the form

$$\nabla^2\Theta + \left[R_i + R_e\eta r^{-3}(1 - \eta)^{-2}\right]L_2 v = P(\partial_t + \mathbf{u} \cdot \nabla)\Theta \tag{2c}$$

and the equations for h and g are obtained through the multiplication of the equation of magnetic induction and of its curl by $\mathbf{r}$:

$$\nabla^2 L_2 h = P_m[\partial_t L_2 h - \mathbf{r} \cdot \nabla \times (\mathbf{u} \times \mathbf{B})], \tag{2d}$$

$$\nabla^2 L_2 g = P_m[\partial_t L_2 g - \mathbf{r} \cdot \nabla \times (\nabla \times (\mathbf{u} \times \mathbf{B}))]. \tag{2e}$$

The Rayleigh numbers R_i and R_e, the Coriolis parameter τ, the Prandtl number P and the magnetic Prandtl number P_m are defined by

$$R_i = \frac{\alpha\gamma\beta d^6}{v\kappa}, \quad R_e = \frac{\alpha\gamma\Delta T d^4}{v\kappa}, \quad \tau = \frac{2\Omega d^2}{v}, \quad P = \frac{v}{\kappa}, \quad P_m = \frac{v}{\lambda}, \tag{3}$$

where λ is the magnetic diffusivity. We assume stress-free boundaries with fixed temperatures:

$$v = \partial_{rr}^2 v = \partial_r(w/r) = \Theta = 0 \quad \text{at } r = r_i \equiv \eta/(1 - \eta) \text{ and at } r = r_o = (1 - \eta)^{-1}. \tag{4a}$$

Throughout this chapter, the case $\eta = 0.4$ will be considered unless indicated otherwise. For the magnetic field electrically insulating boundaries are used such that the poloidal function h must be matched to the function $h^{(e)}$ which describes the potential fields outside the fluid shell:

$$g = h - h^{(e)} = \partial_r(h - h^{(e)}) = 0 \quad \text{at } r = r_i \text{ and } r = r_o. \tag{4b}$$

The numerical integration of Eqs (2a)–(2e) together with boundary conditions (4a) and (4b) proceeds with the pseudo-spectral method as described by Tilgner and Busse (1997) which is based on an expansion of all dependent variables in spherical harmonics for the θ, φ-dependences, that is

$$v = \sum_{l,m} V_l^m(r, t) P_l^m(\cos\theta) \exp(im\varphi) \tag{5}$$

and analogous expressions for the other variables, w, Θ, h and g. P_l^m denotes the associated Legendre functions. For the r-dependence expansions in Chebychev polynomials are used. For further details see also Busse *et al.* (1998).

For most of the computations to be reported in the following thirty-three collocation points in the radial direction and spherical harmonics up to the order 64 have been used.

3. Onset of convection in rotating spherical shells

Even the linear problem of the onset of convection described by Eqs (2a)–(2c) in the limit when all non-linear terms can be neglected is a demanding problem because the critical Rayleigh number depends on the Coriolis number τ, the Prandtl number P, the radius ratio η and on the way in which gravity and temperature distribution of the basic state of pure conduction depend on the radius. For the latter dependences, it is usually assumed that both, gravity and the negative temperature gradient increase in proportion to r corresponding to the case $R_e = 0$ in Eqs (2a)–(2c). For the full sphere, $\eta = 0$, the asymptotic scaling of the problem in the limit of large τ has been derived by Roberts (1968) and the conditions for the onset of the columnar mode of convection have been determined by Busse (1970) on the basis of the annulus model of convection. A complete treatment of the asymptotic linear problem has been given only recently by Jones *et al.* (2000) following an earlier analytical study by Yano (1992) based on the model of a thick cylindrical annulus.

Besides the columnar mode of convection, there exists another mode of onset of convection which becomes preferred at sufficiently small Prandtl numbers (Zhang and Busse, 1987). This equatorially attached type of convection flow can be described as inertial waves which are modified by viscous friction and buoyancy forces (Zhang, 1994). Since most of the numerical analysis discussed in this chapter will be confined to more moderate Prandtl numbers we shall not consider equatorially attached convection any further.

The parameter dependence of the onset of convection becomes more complex as spherical shells of finite radius ratio η are considered. Fortunately, the influence of η is relatively weak if the Coriolis parameter is sufficiently high and η does not approach unity too closely. A radius ratio of $\eta = 0.4$ has been used traditionally (Zhang and Busse, 1987) and extensive computations of the critical conditions for onset have been done (Zhang, 1992a; Ardes *et al.*, 1997). Here, we just want to apply the asymptotic formulas derived for the rotating cylindrical annulus (Busse, 1970, 1986) to give a general impression of the approximate parameter dependence of the onset of convection.

Using the rotating annulus bounded by the coaxial cylinders with radii r_i and r_o as shown in Figure 1 and using the asymptotic relationships (3.9) of Busse (1986) which were derived for the small gap limit we obtain the following expressions for the critical Rayleigh number R_{ic}, the critical wavenumber m_c and the corresponding angular frequency ω_c of wave

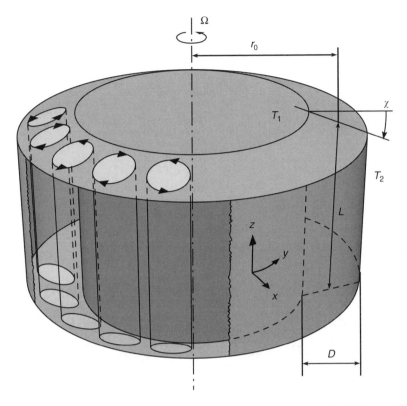

Figure 1 Qualitative sketch of convection columns in the rotating cylindrical annulus.

propagation:

$$R_{\mathrm{ic}} = 3 \left(\frac{P\tau}{1+P} \right)^{4/3} (\tan \theta_{\mathrm{m}})^{8/3} \, r_{\mathrm{m}}^{-1/3} 2^{-2/3}, \tag{6a}$$

$$m_{\mathrm{c}} = \left(\frac{P\tau}{1+P} \right)^{1/3} (r_{\mathrm{m}} \tan \theta_{\mathrm{m}})^{2/3} 2^{-1/6}, \tag{6b}$$

$$\omega_{\mathrm{c}} = \left(\frac{\tau^2}{(1+P)^2 P} \right)^{1/3} 2^{-5/6} (\tan^2 \theta_{\mathrm{m}}/r_{\mathrm{m}})^{2/3}. \tag{6c}$$

Here r_{m} refers to the mean radius of the fluid shell, $r_{\mathrm{m}} = (r_{\mathrm{i}} + r_{\mathrm{o}})/2$, and θ_{m} to the corresponding colatitude, $\theta_{\mathrm{m}} = \arcsin (r_{\mathrm{m}}(1 - \eta))$. The basic temperature gradient of the annulus model has been identified with the r-derivative of the temperature in the equatorial plane at the distance r_{m} from the axis. Similar expressions with the same dependence on τ and P can be obtained when the general case (2c) of the basic temperature distribution is used instead of the special case $R_{\mathrm{e}} = 0$. It is obvious that the expressions (6a)–(6c) cannot be expected to yield more than an order of magnitude estimate for the accurate parameter values characterizing the onset of convection. But even this rough guideline is useful for the orientation in the vast parameter space. Expression (6a) suggests that R_{c} does not change when $P = 0.5$ instead of $P = 1$ is used while

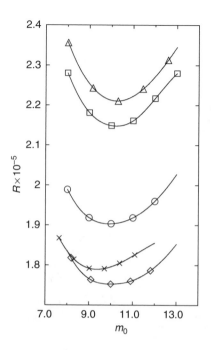

Figure 2 The critical values R_{ic} of the Rayleigh number for the onset of convection in the case $\tau = 10^4$, $P = 1$, $R_e = 0$ for different values of m (indicated by circles). Also shown are values R_{ic} (multiplied by $[P\tau/(1 + P)]^{4/3}$) and corresponding values of m (multiplied by $[P\tau/(1 + \tau)]^{1/3}$) in the cases $\tau = 10^4$, $P = 0.5$ (triangles), $\tau = 10^4$, $P = 2$ (diamonds), $\tau = 1.5 \times 10^4$, $P = 0.5$ (squares) and $\tau = 3 \times 10^4$, $P = 1$ (crosses).

τ is increased by 50%. In Figure 2, examples of this kind have been displayed which indicate that the scaling of expressions (6a)–(6c) is only approximately valid. At least the critical wavenumber m_c appears to obey the scaling Eq. (6b) fairly well as is evident from the nearly identical location of the minima of Figure 2. In the limit of low P the critical Rayleigh number in particular departs rather rapidly from the value given by expression (6a). As shown by Yano (1992) R_c is amplified by a factor of the order P^{-1} from the value (6a) in that limit, owing to the influence of the radial dependence of the convection flow.

The above relationships (6a)–(6c) do not take into account the radial dependence of the convection columns at onset. In fact, it is the property that the onset of convection in the rotating cylindrical annulus becomes independent of the gap width in the asymptotic limit of high τ that make relationships of this kind applicable to spherical shells and other axisymmetric containers. While the annulus model with straight conical end boundaries in the axial direction as shown in Figure 1 yields a phase of the convection columns or thermal Rossby waves independent of radius, the surfaces of constant phase are tilted in the prograde direction with increasing distance from the axis when convexly curved end boundaries are used. The opposite sense of spiralling is found when concavely curved boundaries are applied. This feature is a consequence of the property that the more strongly (weakly) inclined part of the boundary imparts onto the thermal Rossby waves a tendency to propagate faster (slower) which must be compensated by viscous stresses (Busse and Hood, 1982).

The spiralling nature of thermal Rossby waves has an important consequence at finite convection amplitudes: The azimuthally averaged advection of azimuthal momentum does no longer vanish as in the case of a radially independent phase. But instead a differential rotation is generated by the Reynolds stresses of the convection columns. This differential rotation could not be a very strong effect if a feedback process did not come in as well: The steady differential rotation created by the balance of Reynolds and viscous stresses tends to increase the tilt of the convection cells and thus enhances its own source. This feedback mechanism is responsible for the instability that occurs in the case of thermal Rossby waves in the absence of a curved boundary. While this mean flow instability (Busse, 1986; Or and Busse, 1987) can develop in the annulus with either sign, that is, outward transports of prograde as well as of retrograde momentum are possible depending on the sign of the disturbances, only the prograde transport occurs in the case of a spherical shell. The generation of differential rotation by convection in spherical shells could thus be regarded as an imperfect bifurcation of the mean flow instability.

4. Convection in rotating spherical shells at finite amplitudes

The main difficulty faced by the convection flow as it evolves with increasing Rayleigh number is the fact that its columnar structure is not well suited to transport heat from and to the curved boundaries. Since the term describing the transport of heat is basically the same as the term describing the release of potential energy, it is clear that the amplitude of convection cannot grow rapidly with R_i or R_e if the heat transport to the walls is impeded. It is thus not surprising that the amplitude of the steadily drifting convection columns tends to saturate with increasing R as has been demonstrated, for example, in figure 7 of Ardes *et al.* (1997). Convection modes with a strong time dependence tend to replace the steadily drifting columns because of their ability to enhance the heat transport. Most common are vacillating convection columns which either oscillate coherently or incoherently in that a $m = 1$ azimuthal modulation sets in. Vacillating convection can be found in the rotating cylindrical annulus (Or and Busse, 1987), in the spherical case at infinite Prandtl number (Zhang, 1992b), at $P = 1$ (Sun *et al.*, 1993; Grote and Busse, 2001) or at low Prandtl numbers (Ardes *et al.*, 1997). But there are also other ways in which bifurcations from the steadily drifting columns occur, for example through a subharmonic bifurcation as shown in Figure 3. Every second convection column is stretched in the azimuthal direction until the outer tip of the spiraling column snaps off. In the meantime the stretching of the neighboring columns has begun and the process is repeated in a time periodic fashion. If one defines the period of this process as the time in which the same pattern is obtained except for a rotation about the axis, one finds that the period decreases from 0.032 at $R_i = 3.1 \times 10^5$ to 0.021 at $R_i = 3.3 \times 10^5$. At the slightly higher Rayleigh number of $R_i = 3.5 \times 10^5$ a transition to a quasi-periodic time dependence has already occurred which is characterized by a $m = 1 -$ modulation of the pattern. With increasing R_i the convection pattern becomes increasingly chaotic as is evident from the time dependence of the energies of various components of motion shown in Figure 4. The energy of the axisymmetric toroidal component of motion grows especially rapidly with R_i because of the strong differential rotation created by the Reynolds stresses

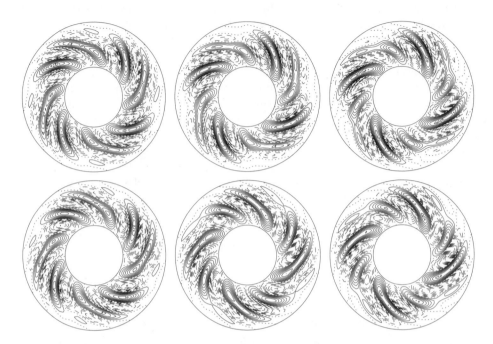

Figure 3 Sequence in time (upper row left to right then lower row right to left) of plots of stream-
lines ($r\,\partial v/\partial\varphi = $ constant) in the equatorial plane in the case $R_i = 3.2\times10^5$, $\tau = 1.5\times10^4$,
$P = 0.5$, $R_e = 0$. The equidistant ($\Delta t = 0.005$) plots cover a period such that the last plots
closely resembles the first plot except for a shift in azimuth.

of the spiralling columns. The kinetic energy can be separated into four different kinds:

$$E_p^m = \tfrac{1}{2}\langle|\nabla \times (\nabla \bar{v} \times \bar{\mathbf{r}}|^2\rangle, \quad E_t^m = \tfrac{1}{2}\langle|\nabla \bar{w} \times \mathbf{r}|^2\rangle, \tag{7a}$$

$$E_p^f = \tfrac{1}{2}\langle|\nabla \times (\nabla \check{v} \times \mathbf{r}|^2\rangle, \quad E_t^f = \tfrac{1}{2}\langle|\nabla \check{w} \times \mathbf{r}|^2\rangle, \tag{7b}$$

where the brackets $\langle\cdots\rangle$ indicate the average over the spherical shell, $\bar{v}$ denote the axisymmetric component of v and $\check{v} \equiv v - \bar{v}$ indicates the non-axisymmetric component of v.
Also presented in Figure 4 are time series of the Nusselt number Nu, which measures the
convective heat transport and which is defined by

$$Nu = 1 - \frac{P}{r_i}\frac{\partial\overline{\overline{\Theta}}}{\partial r}\Bigg|_{r=r_i}, \tag{8}$$

where $\overline{\overline{\Theta}}$ indicates the average of Θ over the surface $r = $ constant. The energy E_t^f has not
been plotted because it closely parallels E_p^f even though it is about twice as large. The energy
of the axisymmetric poloidal component has also not been included because it is several
orders of magnitude smaller than the other energies. As R_i is increased to 5×10^5 the activity
of convection tends to be localized on one side of the sphere similarly as in the case of
$P = 1$, $\tau = 10^4$, $R_i = 7 \times 10^5$ studied by Grote and Busse (2001). Because of the strong
increase in the differential rotation, the radial shear inhibits convection over most parts of
the spherical fluid shell. With increasing R_i, the inhibition of convection by the differential

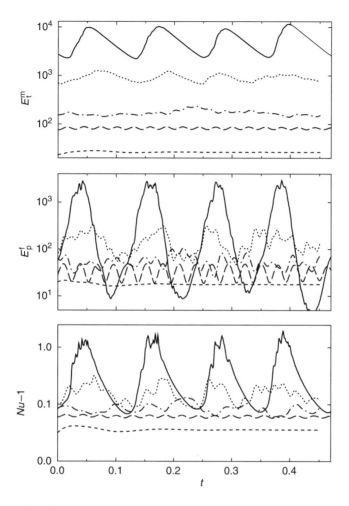

Figure 4 Time series of the mean energy densities of the axisymmetric toroidal component (upper plot) and of the non-axisymmetric poloidal component of motion (middle plot) and of the Nusselt number at the inner boundary (lower plot) in the case $\tau = 1.5 \times 10^4$, $P = 0.5$, $R_e = 0$ for $R_i = 3 \times 10^5$ (short dash lines), 3.5×10^5 (long dash lines), 4×10^5 (dash dotted lines), 6×10^5 (dotted lines) and 10^6 (solid lines).

rotation finally becomes so strong that convection is only possible after the shear has decayed sufficiently in a time interval of near absence of convection. At the end of this interval when the differential rotation reaches its minimum value convection starts to grow for a short moment until the differential rotation has been accelerated by the Reynolds stresses to its previous level and begins to shear off again the convection columns. The resulting relaxation oscillations are characterized by a surprisingly well-defined period in spite of the chaotic nature of the fluctuating part of the velocity field as can be seen from the uppermost lines in each of the three plots of Figure 4. The period of the order of 0.1 corresponds to the viscous decay time of the differential rotating and has been found to be rather independent of P, R_i

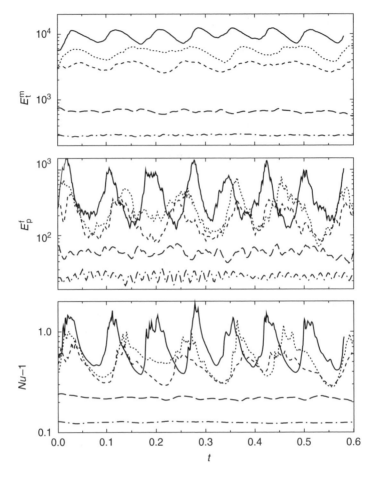

Figure 5 Time series of the mean energy densities of the axisymmetric toroidal component (upper plot) and of the non-axisymmetric poloidal component of motion (middle plot) and of the Nusselt number (lower plot) in the case $\tau = 10^4$, $P = 1$, $R_i = 0$ for $R_e = 6 \times 10^5$ (dash dotted lines), 8×10^5 (long dash lines), 1.2×10^6 (short dash lines), 1.4×10^6 (dotted lines) and 1.7×10^6 (solid lines).

or τ. For other examples of these oscillations, we refer to Grote *et al.* (2000b) and Grote and Busse (2001).

Here we wish to demonstrate that the coherent structures of turbulent convection in rotating spherical shells are not limited to the case of internal heating. When R_e is used as a parameter instead of R_i rather similar phenomena are observed. An overview of the dependence of convection on R_e is provided by Figure 5. After onset of convection a transition to a quasi-periodic form occurs rather quickly with increasing Rayleigh number. In Figure 6, the time dependence of the convection columns is visualized in a sequence of plots. Similar plots for different parameter values are shown by Sun *et al.* (1993). In the absence of internal heating, $R_i = 0$, the temperature gradient near the inner boundary is especially strong and the amplitude of the spiralling convection columns decays more strongly with increasing r than in the case $R_e = 0$. For this reason, the disruption of the convection columns and

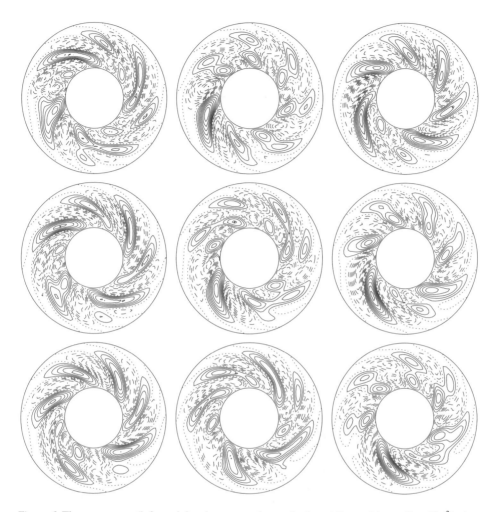

Figure 6 Time sequence (left to right, then top to bottom) of equidistant ($\Delta t = 5 \times 10^{-3}$) plots of streamlines, $r\partial v/\partial \varphi =$ constant in the equatorial plane in the case $\tau = 10^4$, $R_e = 5.5 \times 10^5$, $R_i = 0$, $P = 1$. The kinetic energies exhibit oscillations with the period 0.018, similar to those shown in the case of $R_e = 6 \times 10^5$ in Figure 5.

their subsequent reconnection as shown in the figure occurs more rapidly after onset. At $R_e = 8 \times 10^5$ convection flow has become chaotic as can be seen from the energy of the fluctuating poloidal component of motion. A visualization of the flow at a particular time can be gained from Figure 7. The flow at $R_e = 10^6$ which is also shown in this figure, indicates the tendency towards localized convection although this feature does not seem to develop as dramatically as in the case exhibited in the paper of Grote and Busse (2001). At $R_e = 1.2 \times 10^6$, the relaxation oscillation begins to set in which becomes fully developed as we reach the highest Rayleigh number, $R_e = 1.7 \times 10^6$, of Figure 5. The amplitude of the oscillation of the differential rotation is not quite as large as in the case of Figure 4 and the period is slightly shorter. But the phenomenon itself does not depend on the detailed form of the basic temperature distribution.

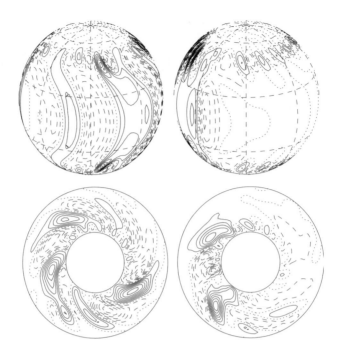

Figure 7 Lines of constant radial component of the velocity on the mid-surface of the shell, $r = (1+\eta)/2(1-\eta)$, (upper plots) and streamlines $(r\partial v/\partial\varphi = \text{constant})$ in the equatorial plane for $R_e = 8 \times 10^5$ (left side) and $R_e = 10^6$ (right side) in the case $\tau = 10^4$, $P = 1$, $R_i = 0$.

5. Convection-driven dynamos

All convection flows realized in sufficiently rapidly rotating spherical shells appear to be capable of acting as dynamos if only the magnetic Prandtl number P_m is sufficiently large, such that a critical magnetic Reynolds number of the order 100 is exceeded even for small amplitudes of convection. A typical diagram is shown in Figure 8, where the open symbols characterize the onset of dynamo action in the R_i–P_m-space for the Coriolis parameter $\tau = 10^4$ and the closed symbols do the same for $\tau = 3 \times 10^4$. In the latter case, the values of R_i must be multiplied by the factor 10 as indicated on the right ordinate. This figure is an extension of a similar one presented by Grote *et al.* (2000b). Analogous diagrams as a function of R_e instead of R_i have been obtained by Christensen *et al.* (1999). There are some typical differences in that dynamos other than dipolar ones are the exception in the cases considered by Christensen *et al.* (see also Kutzner and Christensen, 2000) while in the case of Figure 8 dynamos of predominantly dipolar character are found only for relatively high values of P_m. They are replaced by hemispherical dynamos as P_m decreases and finally at low values of P_m, quadrupolar dynamos are obtained. But the meaning of "high" and "low" values of P_m varies with τ in that the transition from dipolar to hemispherical dynamos shifts towards lower P_m with increasing τ. In the case of $\tau = 3 \times 10^4$ it has not yet been possible to reach the regime of quadrupolar dynamos.

All hemispherical and quadrupolar dynamos exhibit an oscillatory character in that a dynamo wave propagates from the equator to the pole or to the poles, respectively.

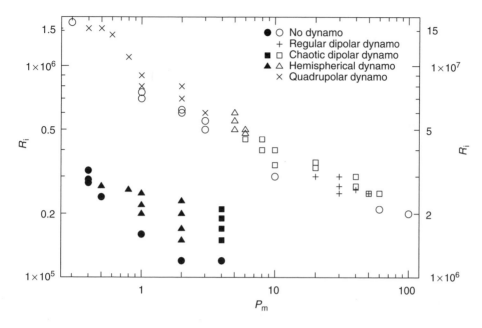

Figure 8 Existence of convection dynamos of different types as a function of the Rayleigh number R_i and the magnetic Prandtl number P_m in the cases $\tau = 10^4$ (open symbols and crosses) and $\tau = 3 \times 10^4$ (filled symbols). In the latter case, the values of R_i are given by the right ordinate. The Prandtl number is $P = 1$ in all cases.

An example is shown in Figure 9. The emergence of magnetic flux with a new polarity appears to be initiated at the equator of the inner boundary as can be best seen in the plots of $\overline{B_\varphi}$ in the figure. The magnetic energy also changes in phase with the oscillation and reaches its maximum when the axisymmetric flux tubes with opposite signs of $\overline{B_\varphi}$ reach about equal amplitude. The perfect correlation between the amplitudes of spherical harmonics with $l - m =$ odd and with $l - m =$ even, which characterizes an ideal hemispherical dynamo is nearly approached in the case $m = 0$ as can be seen from the upper plot of Figure 10. In the case of the non-axisymmetric components, the correlation is not as good, but is still quite remarkable.

A clear distinction of dynamos with different symmetry is usually only possible as long as convection occurs predominantly outside the tangent cylinder and still exhibits an approximate symmetry about the equator. Even in the case of hemispherical dynamos only minor deviation from this symmetry are caused by the action of the Lorentz force. As soon as the critical Rayleigh number for onset of convection in the polar regions is exceeded, the influence of equatorial symmetry diminishes and mixtures of quadrupolar and dipolar components of the magnetic field are generated without the nearly perfect correlation that characterizes hemispherical dynamos. Usually, the quadrupolar part of the magnetic field is still strongest outside the tangent cylinder while the dipolar components appear to be generated primarily in the polar regions. With increasing Rayleigh number convection grows more strongly in the polar regions than outside the tangent cylinder and as a consequence the magnetic field becomes more dipolar. In contrast to the dipolar dynamos indicated in Figure 8, the high

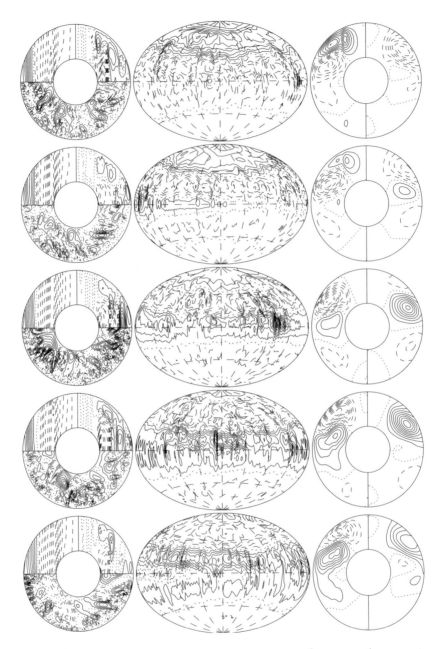

Figure 9 Oscillating hemispherical dynamo in the case $R_i = 2.7 \times 10^6$, $\tau = 3 \times 10^4$, $P_m = 0.5$, $P = 1$. The plots represent a sequence in time (from top to bottom in equidistant steps with $\Delta t = 0.0012$) with the column displaying the velocity field with lines of constant $\overline{u_\varphi}$ in the upper left quarter, streamlines $r \sin\theta \, \partial\overline{v}/\partial\theta = $ constant in the upper right quarter and streamlines $r \partial v/\partial\varphi = $ constant in the lower half of each circle. The middle column shows lines of constant B_r at the outer surface, $r = (1 - \eta)^{-1}$, and right column shows lines of constant $\overline{B_\varphi}$ in the left half and meridional field lines $r \sin\theta \, \partial\overline{h}/\partial\theta = $ constant in the right half of each circle.

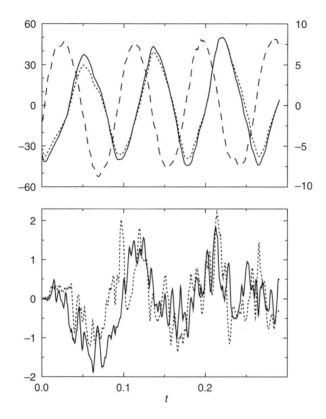

Figure 10 Coefficients $H_l^m(r, t)$ and $G_l^m(r, t)$ (defined for the variables h, g in analogy to definition (5) for v) as a function of time at the position $r = r_i + 0.5$. The upper plot shows $H_1^0(t)$ (dotted line), $H_2^0(t)$ (solid line) and $G_1^0(t)$ (dashed line). The lower plot shows $G_1^1(t)$ (dotted line) and $G_2^1(t)$ (solid line).

Rayleigh number dipolar fields still exhibit the oscillations that characterize the predominantly quadrupolar fields at lower values of R_i. An example of an oscillatory mainly dipolar dynamo is shown in Figure 11.

6. Interaction of magnetic fields and convection

The problem of convection in rotating systems in the presence of an imposed magnetic field has a long history. It can be formulated as a linear problem when the conditions for the onset of convection are of primary interest. Early work on the subject is reviewed in Chandrasekhar's (1961) monograph. Here it has been demonstrated that the stabilizing effect of the Coriolis force on the onset of convection in a horizontal fluid layer heated from below and rotating about a vertical axis can be partly released when a vertical magnetic field is imposed. Even an imposed horizontal field may lead to reduction of the critical Rayleigh number R_c for onset of convection if the rotation rate is high enough (Eltayeb, 1972). For a recent study of this problem which includes the possibility of oscillatory onset we refer to Roberts and Jones (2000). The optimal strength B_0 of the magnetic field for lowering R_c is usually given by

Figure 11 Oscillating dipolar dynamo in the case $R_i = 1.4 \times 10^6$, $\tau = 5 \times 10^3$, $P_m = P = 1$. The same quantities as in Figure 9 are plotted with the time step $\Delta t = 0.009$.

an Elsasser number Λ of the order unity where Λ is defined by

$$\Lambda = B_0^2 / \Omega \varrho \mu \lambda. \tag{9}$$

Here Ω denotes the absolute value of the angular velocity of rotation.

In the rotating cylindrical annulus as shown in Figure 1, an axisymmetric azimuthal magnetic field will also lead to a reduction of R_c if the parameter τ is high enough (Busse, 1976; Busse and Finocchi, 1993). The same result has been obtained for an axisymmetric radial field and for the case of convection in the presence of an azimuthal magnetic field in a rotating sphere (Fearn, 1979a,b). Whenever inhomogeneous magnetic fields are imposed such as a curved azimuthal one in the case of the sphere, the possibility of magnetic instability must be taken into account. We do not want to enter the discussion of this subject here and instead refer to the recent paper by Zhang (1995) and the references cited therein. Except for this latter possibility the optimal field strength for the reduction of R_c corresponds to an Elsasser number of the order unity and the characteristic wavenumber α of convection is reduced to a value of the order π or less. While in the absence of a magnetic field the inhibiting effect of the Coriolis force is counteracted by the viscous friction associated with high wavenumber convection rolls, a second minimum of the Rayleigh number in the $R(\alpha)$-relationship typically develops at low values of α when the Lorentz force becomes sufficiently strong. Viscous dissipation is replaced by ohmic dissipation in this case and convection flows in low Prandtl number liquid metals exhibits features which are typical for convection in fluids with much higher Prandtl numbers (Busse *et al.*, 1997; Petry *et al.*, 1997). Non-linear properties of convection subject to both, the effects of rotation and of an imposed azimuthal magnetic field, have also been studied by Olson and Glatzmaier (1995) and by Zhang (1999) in the case of a spherical shell and by Cardin and Olson (1995) in the case of a rotating cylindrical annulus with the same radius ratio as the liquid outer core of the Earth. Besides the imposed azimuthal magnetic field, the case of an imposed uniform magnetic field parallel to the axis of rotation has received much attention (Sarson *et al.*, 1999; Sakuraba and Kono, 2000). The influence of the onset of convection is minimal in this case since the dependence of columnar convection on the coordinate in the axial direction is rather weak. But a change in the non-linear properties of convection is found in the above mentioned papers when Λ exceeds a value of the order unity.

The interaction of convection with magnetic fields generated by its own dynamo action appears to be quite different from that with an imposed magnetic field. No reduction of the critical Rayleigh number R_c for onset of convection has yet been found in the case of convection-driven dynamos and significant reductions of the characteristic azimuthal wavenumber of columns have not been noticed in the case of the dynamos discussed in Section 5. The major effect of the generated magnetic field on convection in fluids with P of the order unity or less is the braking of the differential rotation. Through the Lorentz force the magnetic field drains energy from the differential rotation and destroys the relaxation oscillations mentioned in Section 4. The radial extent of the convection columns is increased and the transport of heat is enhanced. These effects are clearly demonstrated in Figure 12, where the relaxation oscillation with an increased energy of the differential rotation returns after the magnetic field has decayed as often happens when the Rayleigh number is just below the value needed for sustained dynamo action.

The main question of the interaction between convection and the magnetic field generated by its dynamo action is the question of the mean amplitude of the magnetic field. With this in mind, extensive computations have been performed to produce the dependences on R_i of the energies of the various components of magnetic and velocity fields as shown in Figure 13. Although the energies are obtained from averages over several ohmic or viscous decay times, the influence of the statistical fluctuations cannot be eliminated entirely and the curves shown in the figure are not as smooth as can be expected if computations would be continued over much longer periods in time. While the magnetic field exhibits a pure

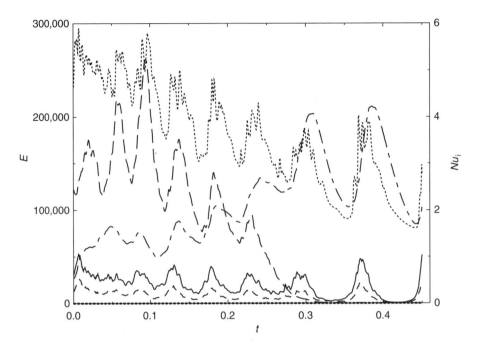

Figure 12 The total magnetic energy (multiplied by 10, dashed line), the energies of the axisymmetric
(dash dotted line) and non-axisymmetric (solid line) toroidal components of motion, $E_t^m V$
and $E_t^f V$, and the non-axisymmetric poloidal component of motion, $E_p^f V$ (dashed line)
are shown as a function of time in the case $\tau = 3 \times 10^4$, $R_i = 2.9 \times 10^6$, $P = 1$, $P_m = 0.4$,
where V denotes the volume of the fluid shell. Also shown is the Nusselt number Nu (right
ordinate, dotted line).

quadrupolar symmetry at low Rayleigh numbers, $R_i \lesssim 5 \times 10^5$, the dipolar component
sets in for $R_i \gtrsim 5 \times 10^5$ which corresponds to the Rayleigh number where convection in
the polar regions becomes possible. There is actually an uncertainty about the onset of the
dipolar component of the magnetic field because it corresponds to a subcritical bifurcation.
As demonstrated in Figure 14, a purely quadrupolar dynamo state as well as one with a small
but finite dipolar component can be realized in the neighborhood of $R_i = 5 \times 10^5$. At
$R_i = 4.8 \times 10^5$ only a purely quadrupolar state is obtained with about 3/4 of the average
energy densities found for $R = 5 \times 10^5$.

   With increasing R_i, the dipolar component grows slightly faster than the quadrupolar
component such that it tends to exceed the latter in energy for $R_i \geq 1.4 \times 10^6$. We have
already mentioned the example of an oscillating predominantly dipolar magnetic field as
shown in Figure 11. The axisymmetric components of both, the dipolar and the quadrupolar
parts of the magnetic field, clearly saturate with increasing R_i. They actually tend to decay
with a further increase of R_i because flux expulsion from the convection eddies becomes
a dominant process. Even the fluctuating components are affected by this process which limits
their energies and leads to a highly filamentary structure of the magnetic field. This latter
property is also evident from the ohmic dissipation which does not exhibit the saturation of
the magnetic energies, but instead parallels the growth of viscous dissipation albeit at a lower
level.

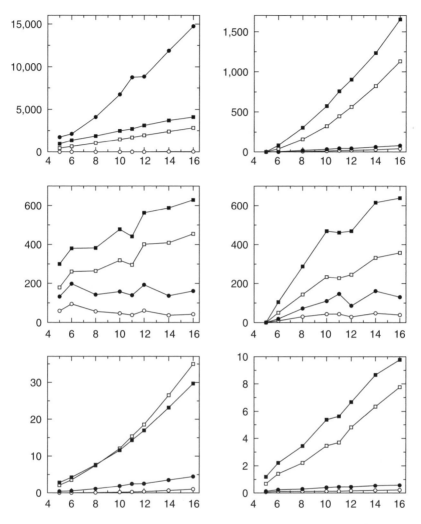

Figure 13 Kinetic energy densities of symmetric (upper left) and anti-symmetric (upper right) components, magnetic energy densities of quadrupolar (middle left) and dipolar (middle right) components, and viscous (lower left) and ohmic (lower right) dissipation are plotted as a function of R_i for convection-driven dynamos in the case $\tau = 5 \times 10^3$, $P = P_m = 1$. Filled (open) symbols indicate toroidal (poloidal) components of the energies and dissipations, circles (squares) indicate axisymmetric (non-axisymmetric) components. In the case of the dissipations, the contributions have not been separated with respect to their equatorial symmetry. The values of R_i at the abscissa should be multiplied by 10^5. The scales of the ordinates in the two lower plots must also be multiplied by the factor 10^5.

Since the Elsasser number can be written in the form:

$$\Lambda = 2 M P_m \tau^{-1}, \tag{10}$$

where M denotes the average density of the total magnetic energy, we conclude from Figure 13 that the value $\Lambda = 1$ is approached by the saturating magnetic field. While such a value makes

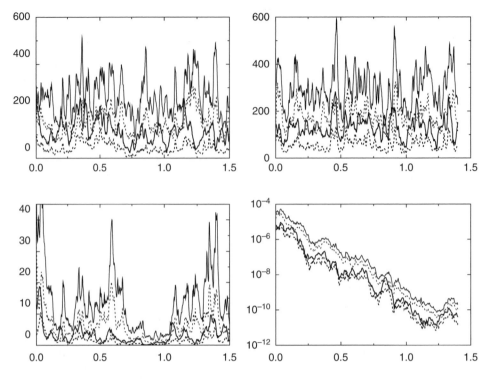

Figure 14 Energies of quadrupolar (dipolar) components of the magnetic field are shown as a function
of time in upper (lower) plots for $\tau = 5 \times 10^3$, $R = 5 \times 10^5$, $P = P_m = 1$. The left side
corresponds to a dynamo of mixed parity, while the right side presents a purely quadrupolar
dynamo. The toroidal energy densities M_t^m, M_t^f (definitions are analogous to those given
by expressions (7a) and (7b)) and poloidal energy densities M_p^m, M_p^f are indicated by solid
and dashed lines, respectively. Thick (thin) lines indicate energy densities of axisymmetric
(non-axisymmetric) components.

sense on the basis of considerations discussed above, the open question remains, why this
value is not reached at lower values of the Rayleigh number. The fact that the ohmic dissipation
amounts to a certain fraction of the viscous dissipation could be a more typical feature of
the interaction between magnetic field and convection in the parameter regime that has been
investigated so far. Further explorations of the parameter dependences of convection-driven
dynamos are clearly needed.

7. The problem of reversals

We have already mentioned that in contrast to dynamos with hemispherical or quadrupolar
structure, dynamos generating dipolar fields do not exhibit oscillatory behavior in general.
Only in the case of high Rayleigh numbers when the dipolar component of the magnetic field
is supported by convection flows in the polar regions can oscillatory behavior be recognized
as shown in Figure 11. Reversals of a predominantly dipolar component of the magnetic field
can be seen, however, as a more randomly occurring event in cases of high Rayleigh number

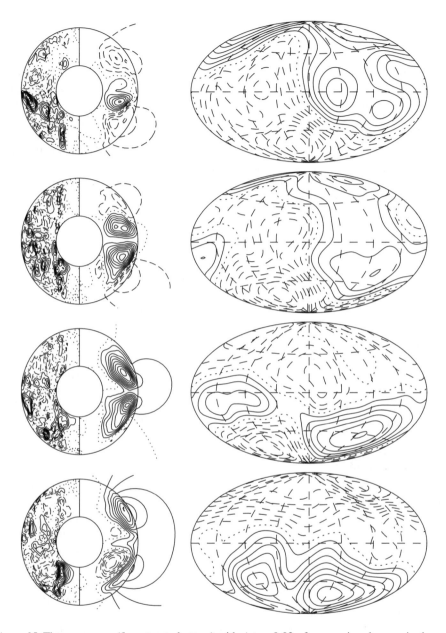

Figure 15 Time sequence (from top to bottom) with $\Delta t = 0.02$ of a reversing dynamo in the case $R = 7 \times 10^5$, $\tau = 10^4$, $P = 1$, $P_m = 6$. The plots on the left side show lines of constant $\overline{B_\varphi}$ in the left half and meridional field lines $r \sin\theta\, \partial \overline{h}/\partial\theta = $ constant in the right half of each circle. The plots on the right side show lines of constant B_r on the surface $r = 2r_0$.

when turbulent convection generates strongly time dependent fields. An example is shown in Figure 15.

Because the magnetic field is already highly filamentary we have plotted the component B_r on the surface of the sphere with twice the radius of the fluid shell. The plots on the right-hand side of Figure 15 thus emphasize the large-scale structures and indicate substantial deviations from the structure of an axial dipole. If we had plotted lines of constant B_r on the surface of the fluid shell, a filamentary structure of the magnetic field would have been apparent similar to that shown in the plots of the middle column in Figure 11.

Although the transition to the reversed polarity does not occur in an oscillatory manner, the same mechanism still seems to be working. The magnetic field with the new polarity first appears near the equator of the inner sphere and propagates outward and to higher latitude. While this process can be seen most clearly in the mean zonal flux, $\overline{B_\varphi}$, in the case of the oscillatory dynamo of Figure 11, it is now more clearly evident in the structure of the meridional field. The flux tubes of $\overline{B_\varphi}$ = constant have assumed such a small scale in Figure 15 that their propagation can no longer be easily seen.

While the present computations have been continued for several viscous diffusion times, longer integration times are needed to obtain a statistically valid sample of reversals. It will also be of interest to study the effect of an electrically conducting inner core on the reversal process in its oscillatory and its more random manifestations. The computations done by Hollerbach and Jones (1993) have indicated a strong influence of the inner core, but our preliminary computations do not indicate a significant effect in the case of reversals with a highly turbulent dynamo.

Occasional reversals in dynamo simulations exhibiting dominant fluctuating components of the magnetic field have been found in previous work (see, e.g. Kida and Kitauchi, 1998) and appear to be a common phenomenon when the axisymmetric components of the magnetic field play a secondary role in comparison to the non-axisymmetric ones. This situation cannot be excluded for the Earth's core and some paleomagnetic observations (Guyodo and Valet, 1999) indicate strongly fluctuating intensity variations. But because the parameter regime accessible to numerical dynamo simulations is still quite different from that of the Earth's core, it is too early to draw definitive conclusions from similarities between computed and observed magnetic fields.

8. Concluding remarks

The dynamo studies discussed in this chapter have been motivated by geophysical applications. But it is still doubtful whether the mechanisms explored hitherto can provide a realistic description of the geodynamo. It has been difficult, for instance, to reach a magnetic Prandtl number much less than unity. It may not be necessary to attain the value 10^{-6} based on the molecular diffusivities expected to characterize the liquid outer core of the Earth. But the effects of small-scale turbulence are unlikely to raise the magnetic Prandtl number much beyond 10^{-2}. Christensen *et al.* (1999) have suggested that the critical value of P_m for dynamo action decreases like $750\tau^{-3/4}$ with increasing τ. We are less optimistic in this respect since our computations suggest a much weaker dependence.

A common property of most computational dynamos is that ohmic dissipation is either less or not far above viscous dissipation. With decreasing Prandtl number the ratio of ohmic to viscous dissipation usually increases as can be seen from figure 7(b) of Grote *et al.* (2000b). It thus seems prudent to proceed in the direction of lower P in order to increase that ratio, especially since low values of P are typical for liquid metals like those in planetary cores. On the other hand, the effective Prandtl number must be large in the Earth's core since large ratios between magnetic and kinetic energies can be expected only in the case of large P (Glatzmaier and Roberts, 1995; Busse *et al.*, 1998). This latter property can be used for arguing that the concentration of light elements with its low diffusivity rather than the temperature provides the buoyancy for driving convection. A combination of both effects may even lead to new effects which have not yet been fully explored.

References

Ardes, M., Busse, F. H. and Wicht, J., "Thermal convection in rotating spherical shells," *Phys. Earth Planet. Inter.* **99**, 55–67 (1997).

Busse, F. H., "Thermal instabilities in rapidly rotating systems," *J. Fluid Mech.* **44**, 441–460 (1970).

Busse, F. H., "Generation of planetary magnetism by convection," *Phys. Earth Planet. Inter.* **12**, 350–358 (1976).

Busse, F. H., "Asymptotic theory of convection in a rotating cylindrical annulus," *J. Fluid Mech.* **173**, 545–556 (1986).

Busse, F. H. and Finocchi, F., "On the onset of thermal convection in a rotating cylindrical annulus in the presence of a magnetic field," *Phys. Earth Planet. Inter.* **80**, 13–23 (1993).

Busse, F. H. and Hood, L. L., "Differential rotation driven by convection in a rotating annulus," *Geophys. Astrophys. Fluid Dynam.* **21**, 59–74 (1982).

Busse, F. H., Clever, R. M. and Petry, M., "Convection in a fluid layer heated from below in the presence of homogeneous magnetic fields," *Acta Astron. Geophys. Univ. Comen.* **19**, 179–194 (1997).

Busse, F. H., Grote, E. and Tilgner, A., "On convection driven dynamos in rotating spherical shells," *Studia Geoph. Geod.* **42**, 211–223 (1998).

Cardin, P. and Olson, P., "The influence of toroidal magnetic field on thermal convection in the core," *Earth Planet. Sci. Lett.* **132**, 167–183 (1995).

Chandrasekhar, S., *Hydrodynamic and Hydromagnetic Stability*. Clarendon Press, Oxford (1961).

Christensen, U., Olson, P. and Glatzmaier, G. A., "Numerical modeling of the geodynamo: a systematic parameter study," *Geophys. J. Int.* **138**, 393–409 (1999).

Eltayeb, I. A., "Hydromagnetic convection in a rapidly rotating fluid layer," *Proc. R. Soc. Lond.* A **326**, 229–254 (1972).

Fearn, D. R., "Thermal and magnetic instabilities in a rapidly rotating fluid sphere," *Geophys. Astrophys. Fluid Dynam.* **14**, 103–126 (1979a).

Fearn, D. R., "Thermally driven hydromagnetic convection in a rapidly rotating sphere," *Proc. R. Soc. Lond.* A **369**, 227–242 (1979b).

Glatzmaier, G. A. and Roberts, P. H., "A three-dimensional convective dynamo solution with rotating and finitely conducting inner core and mantle," *Phys. Earth Planet. Inter.* **91**, 63–75 (1995).

Grote, E. and Busse, F. H., "Dynamics of convection and dynamos in rotating spherical fluid shells," *Fluid Dynam. Res.* **28**, 349–368 (2001).

Grote, E., Busse, F. H. and Tilgner, A., "Convection driven quadrupolar dynamos in rotating spherical shells," *Phys. Rev. E* **60**, R5025–R5028 (1999).

Grote, E., Busse, F. H. and Tilgner, A., "Effects of hyperdiffusivities on dynamo simulations," *Geophys. Res. Lett.* **27**, 2001–2004 (2000a).

Grote, E., Busse, F. H. and Tilgner, A., "Regular and chaotic spherical dynamos," *Phys. Earth Planet. Inter.* **117**, 259–272 (2000b).

Guyodo, Y. and Valet, J.-P., "Global changes in intensity of Earth's magnetic field during the past 800 kyr," *Nature* **399**, 249–252 (1999).

Hollerbach, R. and Jones, C. A., "A geodynamo model incorporating a finitely conducting inner core," *Phys. Earth Planet. Inter.* **75**, 317–327 (1993).

Jones, C. A., Soward, A. M. and Mussa, A. I., "The onset of thermal convection in a rapidly rotating sphere," *J. Fluid Mech.* **405**, 157–179 (2000).

Kida, S. and Kitauchi, H., "Thermally driven MHD dynamo in a rotating spherical shell," *Prog. Theor. Phys. Suppl.* **130**, 121–136 (1998).

Kutzner, C. and Christensen, U., "Effects of driving mechanisms in geodynamo models," *Geophys. Res. Lett.* **27**, 29–32 (2000).

Olson, P. and Glatzmaier, G. A., "Magnetoconvection in a rotating spherical shell: structure of flow in the outer core," *Phys. Earth Planet. Inter.* **92**, 109–118 (1995).

Or, A. C. and Busse, F. H., "Convection in a rotating cylindrical annulus. Part 2. Transitions to asymmetric and vacillating flows," *J. Fluid Mech.* **174**, 313–326 (1987).

Petry, M., Busse, F. H. and Finocchi, F., "Convection in a rotating cylindrical annulus in the presence of a magnetic field," *Eur. J. Mech. B: Fluids* **16**, 817–833 (1997).

Roberts, P. H., "On the thermal instability of a rotating-fluid sphere containing heat sources," *Phil. Trans. R. Soc. Lond.* A **263**, 93–117 (1968).

Roberts, P. H. and Jones, C. A., "The onset of magnetoconvection at large Prandtl number in a rotating layer I. Finite magnetic diffusion," *Geophys. Astrophys. Fluid Dynam.* **92**, 289–325 (2000).

Sakuraba, A. and Kono, M., "Effect of a uniform magnetic field on nonlinear magnetoconvection in a rotating fluid spherical shell," *Geophys. Astrophys. Fluid Dynam.* **92**, 255–287 (2000).

Sarson, G. R., Jones, C. A. and Zhang, K., "Dynamo action in a uniform ambient field," *Phys. Earth Planet. Inter.* **111**, 47–68 (1999).

Sun, Z.-P., Schubert, G. and Glatzmaier, G. A., "Transitions to chaotic thermal convection in a rapidly rotating spherical fluid shell," *Geophys. Astrophys. Fluid Dynam.* **69**, 95–131 (1993).

Tilgner, A. and Busse, F. H., "Finite amplitude convection in rotating spherical fluid shells," *J. Fluid Mech.* **332**, 359–376 (1997).

Yano, Y.-I., "Asymptotic theory of thermal convection in rapidly rotating systems," *J. Fluid Mech.* **243**, 103–131 (1992).

Zhang, K., "Spiralling columnar convection in rapidly rotating spherical fluid shells," *J. Fluid Mech.* **236**, 535–556 (1992a).

Zhang, K., "Convection in a rapidly rotating spherical shell at infinite Prandtl number: transition to vacillating flows," *Phys. Earth Planet. Inter.* **72**, 236–248 (1992b).

Zhang, K., "On coupling between the Poincaré equation and the heat equation," *J. Fluid Mech.* **268**, 211–229 (1994).

Zhang, K.-K., "Spherical shell rotating convection in the presence of a toroidal magnetic field," *Proc. R. Soc. Lond.* A **448**, 245–268 (1995).

Zhang, K., "Nonlinear magnetohydrodynamic convective flows in the Earth's fluid core," *Phys. Earth Planet. Inter.* **111**, 93–103 (1999).

Zhang, K.-K. and Busse, F. H., "On the onset of convection in rotating spherical shells," *Geophys. Astrophys. Fluid Dynam.* **39**, 119–147 (1987).

Zhang, K. and Jones, C. A., "The effect of hyperviscosity on geodynamo models," *Geophys. Res. Lett.* **24**, 2869–2873 (1997).

7 Dynamo and convection experiments

Henri-Claude Nataf

LGIT, Université Joseph-Fourier de Grenoble, BP53, 38041 Grenoble Cedex 9, France

Year 2000 was marked by the success of the dynamo experiments of Riga and Karlsruhe. In both cases, a saturated magnetic field of a few milliTeslas was produced. These successes crown years of efforts of the two teams during which the experiments were planned, built and tested. In both cases, more than a cubic meter of liquid sodium is used and the flow is driven into a well-defined configuration with powers in excess of 100 kW. The configurations were chosen to mimic known kinematic dynamos, thus enabling a good prediction of the onset of dynamo action. These experiments demonstrate the feasibility of laboratory dynamos and open the way to a second generation of experiments, in which the flow will have more freedom to organize itself under the combined actions of the forcing and of the Lorentz force. It will also be interesting to explore the effect of the Coriolis force and to unravel the characteristics of turbulence in these dynamos. All these dynamo experiments nicely complement the numerical models now available because they allow the exploration of regimes with low magnetic Prandtl number and high Reynolds number.

I also review laboratory experiments that aim at the understanding of convection in planetary cores. There also, turbulent regimes are more easily reached than in numerical experiments, and new features have been discovered recently. Crystallization experiments explore a new class of mechanisms with implications for the anisotropy of the inner core.

Experiments continue to play an important role in the modelling of mantle convection, as illustrated by the recent findings on the organization of convection in two superposed layers. New types of behaviours are being investigated, with emphasis on nonlinear physics and dynamical systems.

1. Introduction

This chapter focuses on laboratory experiments that have been carried out recently to help understand the dynamics of the Earth's deep interior. Experiments have always played a major role in fluid mechanics. Before computers became available, experiments were the main providers of ideas and data. Now that numerical modelling has become the powerful tool that we know, experiments continue to play a unique role in fluid mechanics.

They often complement the parameter space available for numerical modelling. For example, we will see that the very low values of the magnetic Prandtl number P_m ($P_m = \nu/\eta$, where ν is the kinematic viscosity and η the magnetic diffusivity) expected for liquid iron in the Earth's core cannot be reached in numerical models, while they are typical of liquid metals such as gallium and sodium that are used in laboratory experiments.

Another nice thing about experiments is that they force the realization of physically realistic situations. If a magnetic field is imposed, it has to be produced by a magnet or by a coil in some physical way. Boundaries, which often play a key role in the dynamics of rotating fluids, have to be present as well in an actual experiment.

Even more interesting is the fact that experiments often give another viewpoint. The problem of precession, discussed by Aldridge in Chapter 8 of this volume provides a striking illustration of this remark: while flow in a precessing sphere seems dominated by shear on geostrophic cylinders in laboratory experiments, the most conspicuous features in numerical models are inclined bands of oscillatory motions. It is only very recently that the two viewpoints have been shown to be two facets of the same complex phenomenon. The common sense intuition is indeed often defeated when dealing with rotating fluids, and even more when a magnetic field is present. Experiments certainly help improve our physical appraisal of what is going on.

While the explosion of computer resources clearly plays a central role in the fascinating progression of numerical modelling, it is not always realized that, together with other technological improvements, they have also considerably enriched the output of experiments. New measuring techniques, such as particle image velocimetry (PIV) or ultrasonic Doppler velocimetry, are available and large quantities of data can now be handled and processed easily in the lab. New technologies have also appeared. For example, the use of liquid sodium in the nuclear industry has led to a much better knowledge of its properties and handling requirements, and this is very valuable for scientists that design dynamo experiments today.

Finally, I note that the widespread availability of computer resources and numerical codes makes it possible for a small group of investigators to carry on experiments and numerical simulations together. Several teams in the world operate in this fashion, and this appears to be very fruitful.

The event of year 2000 was the success of two dynamo experiments, and the first section of this chapter is devoted to these experiments. The main results are reviewed, with emphasis on the new questions that arise, opening the way to a new generation of dynamo experiments. I then discuss recent experimental findings concerning convection in the Earth's core – including crystallization – and in the Earth's mantle.

2. Dynamo experiments

Following the ideas of Larmor (1919), Elsasser (1946) and Bullard and Gellman (1954), it is now widely accepted that the magnetic field of the Earth is sustained by some sort of dynamo mechanism in its metallic liquid iron core (see Busse *et al.* in Chapter 6 of this volume). Consider the induction equation in a continuum:

$$\frac{\partial \mathbf{B}}{\partial t} + \nabla \times (\mathbf{B} \times \mathbf{u}) = \eta \nabla^2 \mathbf{B}. \tag{1}$$

For suitable imposed velocity fields $\mathbf{u}$, the solution $\mathbf{B} = \mathbf{0}$ is unstable and a magnetic field appears and grows when the magnetic Reynolds number $R_m = UD/\eta$ is large enough, where U is a typical velocity and D a typical length. For a given velocity field, it is possible to evaluate the critical magnetic Reynolds number R_{m_c} of instability: this is the basis of the kinematic dynamo approach.

One of the first tasks of the proponents of the dynamo mechanism was to find velocity fields that could act as a dynamo. Among the various solutions found, three have now found an experimental counterpart (see Lielausis, 1994 and Tilgner, 2000 for recent reviews on experimental dynamos).

Table 1 Physical properties of a few liquids used in the experiments

Property	Symbol	Unit	Sodium, 200°C	Gallium, 30°C	Water, 20°C
Density	ρ	$kg\,m^{-3}$	910	6,095	1,000
Kinematic viscosity	ν	$m^2\,s^{-1}$	5×10^{-7}	2.95×10^{-7}	10^{-6}
Coefficient of thermal expansion	α	K^{-1}	2.9×10^{-4}	1.26×10^{-4}	2×10^{-4}
Thermal conductivity	k	$W\,m^{-1}\,K^{-1}$	82	30	0.59
Heat capacity	C_p	$J\,K^{-1}\,kg^{-1}$	1,330	381	4,180
Thermal diffusivity	κ	$m^2\,s^{-1}$	0.66×10^{-4}	1.3×10^{-5}	1.4×10^{-7}
Electric conductivity	σ	$\Omega^{-1}\,m^{-1}$	7.5×10^6	3.87×10^6	–
Magnetic diffusivity (at 150°C)	η	$m^2\,s^{-1}$	0.106 (0.092)	0.21	–
	$\rho\eta^3$	$W\,m$	1.08	53	–
Prandtl number	Pr	ν/κ	7.5×10^{-3}	0.023	7.1
Magnetic Prandtl number	P_m	ν/η	4.7×10^{-6}	1.4×10^{-6}	–

The first experimental dynamo was built by Lowes and Wilkinson (1963, 1968). It was inspired from the kinematic dynamo of Herzenberg (1958) and nicely depicted the phenomenon of reversals. Two solid cylinders were rotated around nonparallel axes. Mercury provided electrical continuity between the two cylinders. The device did work as a dynamo because the cylinders were made of ferromagnetic iron, with a very low magnetic diffusivity, thus enabling large magnetic Reynolds numbers to be reached for moderate size and velocity.

The two other kinematic dynamos that have inspired actual experiments are those of Ponomarenko (1973) and Roberts (1972). They provide a theoretical guide to the experiments of Riga and Karlsruhe, respectively. These two experiments are homogeneous liquid dynamos and no ferromagnetic part is present.

All the experimental dynamos described below use liquid sodium as a working liquid, because of its low magnetic diffusivity and low density. The relevant physical properties of liquid sodium are given in Table 1. Note that the magnetic diffusivity of liquid sodium increases with temperature. Hence, dynamo action is more easily reached at low temperature (but freezing of sodium – at 98°C – is to be avoided!).

2.1. The Riga dynamo experiment

The success of the Riga dynamo in year 2000 crowns 30 years of efforts under the guidance of Professor Agris Gailitis. The set-up of the dynamo, shown in Figure 1, was designed by the Riga group to mimic the Ponomarenko (1973) kinematic dynamo. The dynamo module is a 3 m long cylinder, containing about 1.5 m³ of liquid sodium, split into three co-axial cylindrical volumes. In the inner cylinder (0.25 m in diameter), liquid sodium flows down in a helicoidal fashion, forced by a propeller at the top. The return flow takes place vertically between the inner tube and the surrounding one. The outermost shell of sodium is at rest. The propeller is driven at up to 2,200 rpm by two 55 kW electric motors. The expected eigenmode for the magnetic field consists in a vector field that also follows an helice and oscillates in time at a given position. The critical magnetic Reynolds number was determined by Ponomarenko (1973) for an infinitely long cylinder. Gailitis and Freibergs (1976) computed the threshold in a finite cylinder for both the convective and the absolute instability.

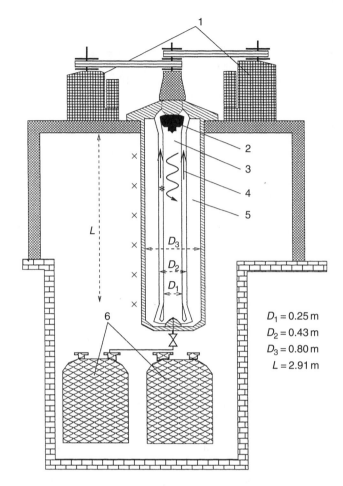

Figure 1 The Riga dynamo facility. Main parts are: 1 – two motors (55 kW each), 2 – propeller, 3 – helicoidal flow region, 4 – back-flow region, 5 – sodium at rest, 6 – sodium storage tanks, ∗ – position of the flux–gate sensor and the induction coil, × – Positions of the Hall sensors. (Reprinted with permission from Gailitis, A. *et al.*, "Detection of a flow induced magnetic field eigenmode in the Riga dynamo facility," *Phys. Rev. Lett.* **84**, 4365–4368 (2000). Copyright (2000) by the American Physical Society.)

   In 1987, measurements of the decay rate of an applied magnetic field showed that the predicted threshold was correct and could be reached in the experiment (Gailitis *et al.*, 1987). However, mechanical vibrations in the set-up forced the experiment to a stop.

   The module was built anew and the propeller redesigned to drive an optimal flow, and in November 1999, the Riga group observed the slow growth of a magnetic eigenmode for a few seconds before the experiment had to be stopped again because of a sealing problem (Gailitis *et al.*, 2000). Figure 2 shows the actual record of the magnetic field. A 1-s period field was applied, and the beating observed in the record indicates that a signal with a 1.3-s period is slowly growing. This phenomenon was observed for a rotation rate of about 2,150 rpm and

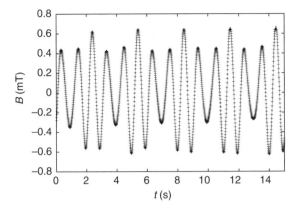

Figure 2 Growth of a magnetic eigenmode of the Riga dynamo. The magnetic field measured by the flux-gate sensor (see Figure 1) in November 1999 for a rotation rate of 2,150 rpm. Note that the imposed 1 Hz signal beats with a growing 1.3 Hz eigenmode. (Reprinted with permission from Gailitis, A. *et al.*, "Detection of a flow induced magnetic field eigenmode in the Riga dynamo facility," *Phys. Rev. Lett.* **84**, 4365–4368 (2000). Copyright (2000) by the American Physical Society.)

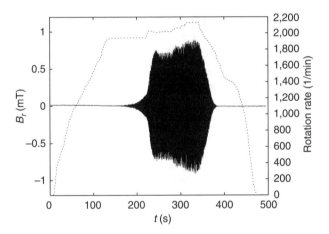

Figure 3 Saturation of the magnetic field in the Riga dynamo. The magnetic field is the continuous line with the scale at the left, while the dotted line gives the variation of the rotation rate (scale on the right) as a function of time during the experiment. Note that as the rotation rate approaches 1,950 rpm, an alternating magnetic field starts growing. At saturation, its amplitude follows the small variations of the rotation rate and then drops when the latter is reduced below the threshold. (Reprinted with permission from Gailitis, A. *et al.*, "Magnetic field saturation in the Riga dynamo experiment," *Phys. Rev. Lett.* **86**, 3024–3027 (2001). Copyright (2001) by the American Physical Society.)

when the temperature of the sodium was about 200°C. The growth rate and the frequency of the magnetic eigenmode are in good agreement with the theoretical predictions.

In July 2000, after the seal was replaced, a series of new runs were conducted with sodium at a temperature of about 160°C. This time, as shown in Figure 3, the magnetic eigenmode

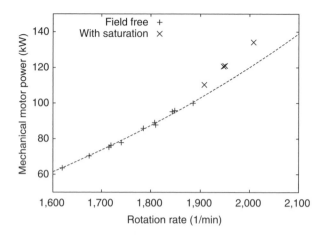

Figure 4   Motor power (in kW) as a function of the rotation rate of the propeller in the Riga experiment,
below (+) and in (×) the dynamo regime. Power scales as rotation rate to the third power
(dashed line) below the dynamo regime. (Reprinted with permission from Gailitis, A. *et al.*,
"Magnetic field saturation in the Riga dynamo experiment," *Phys. Rev. Lett.* **86**, 3024–3027
(2001). Copyright (2001) by the American Physical Society.)

could grow up to saturation levels of a few milliTeslas, and remain at that level for several
minutes (Gailitis *et al.*, 2001). As expected, the power increases as rotation rate to the third
power below the onset of the dynamo. Once the magnetic field is present, the power needed to
sustain a given rotation rate is increased by up to 10 kW, and dissipates in Joule heating (see
Figure 4). Measurements of the magnetic field in this regime reveal that the flow is modified
by the presence of the magnetic field: velocities decrease from the top to the bottom of the
apparatus. *In situ* velocity measurements would help corroborating this observation.

2.2. The Karlsruhe dynamo experiment

The Karlsruhe experiment (Busse *et al.*, 1998a) was built to model another kinematic dynamo:
the two-scale periodic dynamo of Roberts (1972). Figure 5 gives a schematic view of the
actual dynamo module. A 1.7 m diameter, 0.85 m high cylinder is paved with fifty-two 'spin-
generators'. Each of these spin-generators consists of two parts: a 10 cm diameter inner
tube inside a 21 cm diameter outer tube with helicoidal blades. Liquid sodium is driven in
both the inner and outer tube circuits with separate external magnetohydrodynamic pumps at
volumetric rates of about $110 \, \mathrm{m^3 \, h^{-1}}$. The gaps between generators are filled with sodium at
rest. Roberts showed that an infinite pattern of such bidimensional periodic flow could create
a large-scale magnetic field. The threshold for dynamo action was calculated for the finite
configuration by Tilgner (1997) and Rädler *et al.* (1998). The magnetic field fastest growing
eigenmode is predominantly horizontal and rotates by π from the top to the bottom of the
dynamo module.

In the meantime, the dynamo module was built by the group of Professor Müller and
Stieglitz in the Institute for Nuclear and Energy Technologies in Karlsruhe (see Figure 6).

In December 1999, the dynamo was run and a magnetic field was indeed produced (Stieglitz
and Müller, 2001). A steady field of intensity up to 16 mT was observed. The field could be
maintained for more than 10 min. The onset of dynamo action was obtained for flow rates

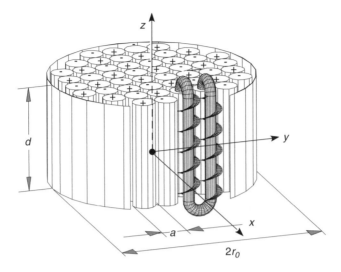

Figure 5 Sketch of the dynamo module of the Karlsruhe experiment. Fifty-two spin-generators are packed inside a 1.7 m diameter cylinder. The inner structure is shown for two spin-generators in the front row. The signs indicate the direction of the sodium flow in individual spin-generators. (Reprinted with permission from Stieglitz, R. and Müller, U., "Experimental demonstration of a homogeneous two-scale dynamo," *Phys. Fluids* **13**, 561–564 (2001). Copyright (2001) by the American Physical Society.)

Figure 6 Photograph of the dynamo module of the Karlsruhe experiment. (Courtesy of Professor Müller.)

slightly above $100 \, \text{m}^3 \, \text{h}^{-1}$ with liquid sodium at a temperature of 125°C. The threshold thus appears to be somewhat lower than in the theoretical predictions (Figure 7). In the year 2000, more runs were performed. The geometry of the magnetic field was found to be in good agreement with the predictions, but could depend on the initial electromagnetic conditions. More attention was given to the saturation mechanism. One could have thought that once the

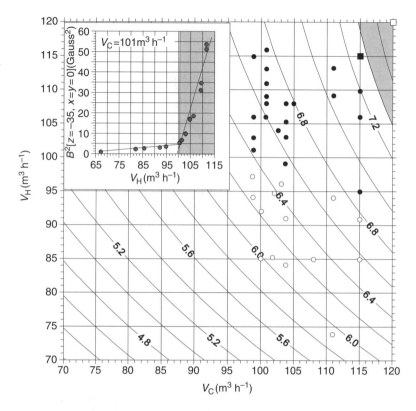

Figure 7 Stability diagram for the onset of dynamo action in the Karlsruhe experiment. V_C is the flow rate in the central part of the spin-generators and V_H is the flow rate in their helicoidal outer part (see Figure 5). The filled circles are for runs where dynamo action was observed, while open circles denote nondynamo states. The grey area specifies the dynamo domain according to Rädler *et al.*'s predictions. The two squares indicate Tilgner's predictions for dynamo onset. The inset gives the evolution of the magnetic energy as a function of V_H when V_C is held fixed. Note the sharp increase that marks the dynamo onset (grey domain). (Reprinted with permission from Stieglitz, R. and Müller, U., "Experimental demonstration of a homogeneous two-scale dynamo," *Phys. Fluids* **13**, 561–564 (2001). Copyright (2001) by the American Physical Society.)

magnetic field has appeared, an increase in the pressure head in the pumps will only increase the magnetic field and leave the velocity field unchanged (Busse *et al.*, 1998a). Instead, it appears that the flow rate continue to increase above the dynamo threshold. Tilgner and Busse (2001) interpret this as an indication that the Lorentz force, which grows as the magnitude of the magnetic field increases, does not simply reduce the amplitude of the liquid sodium flow but also modifies the velocity field in some way. *In situ* velocity measurements would help understanding this phenomenon.

It is clear that the success of the dynamo experiments in Riga and Karlsruhe represent a key milestone. It provides an experimental validation of the kinematic dynamo approach. It shows that such experiments are feasible, as impressive and difficult as they might look. In the two experiments, the flow is turbulent since the Reynolds number reaches several millions. However, the mean flow is much larger than its fluctuations. The Riga experiment demonstrates that, at the attained saturation level of the magnetic field, the Lorentz force does not modify the mean flow drastically. Clearly, it would be interesting to monitor the velocity field in this situation. The Karlsruhe experiment demonstrates the efficiency of the separation of scales (small-scale velocity field creating a large-scale magnetic field). The two configurations were chosen because a magnetic field was predicted to appear for 'low' values of the magnetic Reynolds number. Nevertheless, both experiments need more than a cubic meter of sodium and dissipate powers of more than 100 kW. Future dynamo experiments will probably have similar requirements.

2.3. Turbulent dynamos

Indeed, a second generation of experimental dynamos is under way. The force of the 'kinematic' dynamo approach of Riga and Karlsruhe was to yield robust predictions for the dynamo threshold. Its main weakness is that the Lorentz force is not allowed to modify the mean flow in a strong way. This has lead several teams to propose new dynamo set-ups, in which the forcing is much less geometrically constrained.

2.3.1. Convective dynamos?

A typical example would be a convective dynamo: convective motions appear spontaneously once a sufficient temperature gradient is imposed, and when the convective velocities are large enough, a magnetic field can be created and is able to modify the geometry of the convective cells. This approach was followed with great success by numerical modellers in the recent years, but for hypothetical materials with physical properties far from those of liquid metals (see Busse *et al.* in Chapter 6 of this volume, and recent reviews by Fearn (1998) and Roberts and Glatzmaier (2000)). A convective dynamo is probably not feasible in the lab. If we extrapolate the scaling laws recently derived for the convective velocities in a rotating sphere filled with liquid gallium (see below), we get that typical convective velocities would be of only $0.1 \, \mathrm{m \, s^{-1}}$ in a 1 m radius sphere rotating at 200 rpm and filled with liquid sodium when a thermal power of 100 kW is driving convection. Such velocities are probably two orders of magnitude too low to enable the growth of a magnetic field. After the early experiments of Dan Lathrop at the University of Maryland (Peffley *et al.*, 2000b), a similar conclusion was reached by Glatzmaier (2000). Nevertheless, Lathrop is presently conducting a 'dynamo' convection experiment in a 60 cm diameter sphere filled with liquid sodium, spinning at up to 6,000 rpm, with a driving thermal power of 10 kW. The challenge remains tantalizing, especially since doubts have been cast on the possibility of convective dynamo action in a low Prandtl number liquid (Busse *et al.*, 1998b).

2.3.2. Mechanical dynamos

Most investigators (including Lathrop) pursue another, more reachable, goal. The idea is to force fluid motions in a more efficient way by spinning some propeller in a tank filled with liquid sodium. Several features of the convective dynamo would be retained. In particular, the Lorentz force could strongly modify both the mean flow and its fluctuations. For dynamo action to take place, the magnetic Reynolds number must be high (of the order of 50?), hence the usual Reynolds number must be very high (larger than 10^7). As a consequence, it becomes very difficult to predict the onset of dynamo action. Indeed, only simple and stationary flows

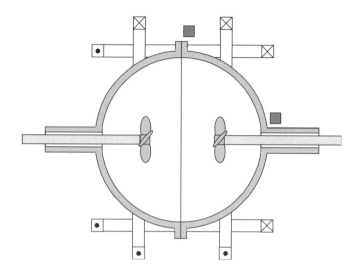

Figure 8 Schematic cross-section of the 'dynamo II' experiment at the University of Maryland. A
30 cm diameter sphere is filled with liquid sodium. Two propellers apply forcing and baffles
enhance the poloidal flow. Also drawn are the magnetic coils that supply external magnetic
fields and the two positions of the magnetometer. (Reprinted with permission from Peffley,
N. L. *et al.*, "Toward a self-generating magnetic dynamo: the role of turbulence," *Phys. Rev.*
E **61**, 5287–5294 (2000). Copyright (2000) by the American Physical Society.)

are easily tested with the kinematic dynamo approach, and the onset of dynamo action is then
found to be very sensitive to details of the imposed velocity field. Furthermore, in contrast to
the onset of convection in most situations, an increase in the control parameter (the amplitude
of velocity) can lead to a decrease of the growth rate of a seed magnetic field. Entering the
dynamo regime in this configuration could therefore be very different from what is seen in
the dynamos of Riga and Karlsruhe. In fact, there is no guarantee that a given forcing will
ever produce a dynamo!

Since it is difficult to predict the dynamo onset of a turbulent dynamo using the kinematic
dynamo approach, some efforts have been put in determining it experimentally in small size
experiments. The approach was pioneered by Lathrop (Peffley *et al.*, 2000a). Figure 8 shows
the set-up he built. Two propellers along the same polar axis spin in opposite senses inside a
30 cm diameter sphere filled with liquid sodium. One expects a flow with a strong azimuthal
component and a meridional circulation, both antisymmetric with respect to the equator. One
of the most interesting results obtained by Lathrop is shown in Figure 9. In order to estimate
how far from the onset he was, he monitored the growth rate of a pulse of magnetic field,
in a similar way to Gailitis *et al.* (1987). When the propellers are at rest, the magnetic field
growth rate is negative and its amplitude is given by the inverse of the magnetic diffusion
time. When the propellers spin, the advection of the magnetic field by the flow can lead to
an increase or a decrease of the growth rate. The dynamo onset corresponds to a zero growth
rate. In addition, the growth rate depends on the geometry of the magnetic field. Lathrop
tested two different geometries: an axisymmetric ($m = 0$) field and a $m = 1$ field pointing
in a direction perpendicular to the rotation axis. While one could expect the latter to be the
fastest growing mode, based on the similitude of the mean flow with axisymmetric $s_2 + t_2$
flows tested by Dudley and James (1989), the measurements indicate that, on the contrary, the

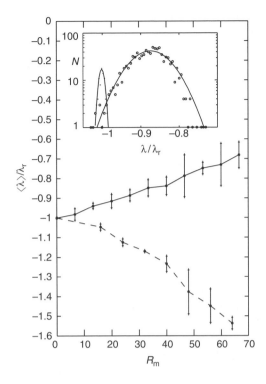

Figure 9 Growth rate λ of the magnetic field as a function of magnetic Reynolds number in the 'dynamo II' experiment of the University of Maryland. It shows a shift toward self-generation for fields aligned with the shafts ($m = 0$) (solid line) and a shift away from self-generation for fields at right angle ($m = 1$) (dashed line). The error bars indicate shot-to-shot fluctuations in the growth rate estimates. These fluctuations are further quantified by the distribution shown in the inset for 895 observations of the growth rate at the same $R_{\rm m} = 42$. All the growth rates in this figure are normalized to yield -1 in conditions of no flow. The distribution for no flow $R_{\rm m} = 0$ is the narrow Gaussian around -1 in the inset. Presumably, self-generation state is achieved when the $λ = 0$ line is crossed. (Reprinted with permission from Peffley, N. L. *et al.*, "Toward a self-generating magnetic dynamo: the role of turbulence," *Phys. Rev. E* **61**, 5287–5294 (2000). Copyright (2000) by the American Physical Society.)

$m = 1$ field decays more rapidly when the velocity increases. In contrast, the axisymmetric mode (which would always decay for a purely axisymmetric flow, under Cowling's (1934) theorem) decays less rapidly when motion is present. The inset also shows that in a turbulent flow, the distribution of instantaneous growth rates is very wide.

2.3.3. *Small dynamos?*

Several teams count on 'small' size experiments to reach the dynamo regime. Indeed, since the Reynolds number is linear in both size and velocity, one can choose to increase flow velocity rather than size, especially if one bears in mind that the volume of the experiment varies as the size to the power three. Following this idea, dynamo experiments with sodium volumes of less than 100 l have been designed and run by Lathrop (Peffley *et al.*, 2000a) and by the French VKS group in Cadarache (Bourgoin *et al.*, 2001). The approach is also followed by the New Mexico group with 150 l (Colgate *et al.*, 2001a,b). These experiments are very interesting but they have not produced a dynamo yet. A simple argument helps understanding why. In the turbulent regime, the power needed to drive a flow scales as:

$$P \sim \rho \eta^3 \frac{R_m^3}{D}. \tag{2}$$

For a given flow, in order to reach the R_m needed for dynamo onset, a smaller experiment will need a higher driving power, hence a much higher power density. The difficulty is not so much in providing the power but rather in getting the heat out. It is indeed necessary to remove heat because otherwise the temperature would increase and the conductivity of sodium decrease, thus inhibiting dynamo action. With a volume of 70 l, optimized propellers and a driving power of 150 kW, the VKS group could not make a dynamo (Bourgoin *et al.*, 2001). Indeed, one may infer from their results that a driving power of 300 kW is needed for dynamo onset. Similar values are also obtained when extrapolating the growth rate curve of Lathrop in Figure 9. In other words, there is no such thing as a 'small dynamo', unless one is ready to call a Formula 1 race car a small car. Note that the $\rho \eta^3$ factor in Eq. (2) explains why all dynamo experiments use liquid sodium. This factor is about 1 W m for sodium (see Table 1) while it reaches 50 for gallium and 10,000 for mercury!

With 500 l of liquid sodium and a power of 200 kW, Forest (1998) of the University of Wisconsin at Madison has a chance to observe the dynamo onset once his set-up is operational. Another approach is followed by the Perm group, who investigates the onset of dynamo action in a transient flow (Frick *et al.*, 2000). The idea is to store a large kinetic energy in a 1 m diameter torus filled with 100 l of liquid sodium by spinning it at velocities up to 3,000 rpm. By bringing the torus to a stop very rapidly (in less than 0.1 s), one obtains very large flow velocities and a large transient power release, which might give rise to transient dynamo action.

It is clear that the involvement of physicists in dynamo experiments is rapidly increasing. This is rather natural if one considers the many fundamental issues still to be addressed concerning dynamo action. The expertise on handling liquid sodium acquired by the nuclear industry in the past 30 years is of great help in this enterprise. The experiments now underway have already brought valuable contributions; they have also prompted new questions. For example, how appropriate is the kinematic dynamo approach for predicting the onset of dynamo action in a turbulent flow? Will the onset correspond to a simple direct bifurcation as observed in the Riga and Karlsruhe dynamos, or be intermittent or super-critical (Sweet *et al.*, 2001)?

2.3.4. Geophysical dynamos?

While the geophysical community is clearly playing a leading role in numerical dynamos, its involvement in experimental dynamos is very limited. However, there are a number of issues for which dynamo experiments should prove fundamental. Many of these involve the effect of the Coriolis force. For example, will the Coriolis force help in getting dynamo action? This is a worthy hypothesis, considering that the onset of convection is lowered when both a magnetic field and rotation are present (see later). Another hint stems from the observation that heat transfer is enhanced in numerical models of convective dynamos when the magnetic field is present (Christensen *et al.*, 1999; Grote *et al.*, 2000). Other questions related to the dipolar geometry of the magnetic field and to possible inversions are also of great interest for the geophysical community. Finally, the organization of turbulence in a turbulent rotating dynamo is central to the extrapolation of current numerical models to core conditions, while the power consumption of such a dynamo is important for models of the thermal history of the Earth.

These questions prompted our team in Grenoble to propose a set-up of a 'magnetostrophic dynamo' (Jault *et al.*, 1998). While the project is described elsewhere in more details (Cardin *et al.*, 2002), I just indicate that the idea is to spin a sphere filled with some 4 m^3 of liquid sodium and to force some kind of internal motions by rotating an inner sphere or some other device. For a large enough power input (250 kW as a first estimate), it is expected that a strong magnetic field will appear, such that the Coriolis force will balance the Lorentz force. Most of the dissipation will take place by Joule heating in this regime. Again, since the flow will be very turbulent it is not easy to predict the onset of dynamo for a given forcing. And it is even more difficult to guess what flow will be produced once the magnetic field is acting. For these reasons, as well as for testing several technological issues, a smaller scale experiment has been constructed. A 60 cm diameter sphere filled with 40 l of liquid sodium is entrained by a 10 kW motor to rotation rates up to 2,000 rpm. A 20 cm diameter inner sphere is entrained by another 10 kW motor at a different rotation rate. The inner sphere is made of a permanent magnet, yielding a magnetic field of more than 0.1 T in the liquid, so that the organization of the flow in the presence of both rotation and magnetic field can be investigated.

3. Core convection experiments

We have seen that it was unlikely that a convective dynamo could be produced in the lab. On the other hand, numerical models of convective dynamos now exist (see Busse *et al.* in Chapter 6 of this volume) but for liquids with properties that are very different from what we expect for the Earth's core. In particular, it is very difficult to model liquids with a low Prandtl number $Pr = v/\kappa$ or a low magnetic Prandtl number $P_m = v/\eta$. The organization of convective flows and dynamo action could be very different in this parameter regime. In fact, it has been suggested that dynamo action was not possible for low Prandtl numbers (Busse *et al.*, 1998b). It is therefore important to investigate the properties of convection for a wide range of parameters and try to understand better its characteristics, in the presence of rotation and with a magnetic field, even if the magnetic field is not produced internally by dynamo action. Several interesting results have been recently obtained along these lines.

3.1. Rayleigh–Bénard convection with rotation and a magnetic field

Chandrasekhar's (1961) famous monograph on hydrodynamic and hydromagnetic stability still serves as a reference on these questions. Of particular interest is the stability of an electrically conductive liquid layer between two horizontal plates at different temperatures, subject to rotation around a vertical axis and to a magnetic field in the same direction. When acting separately, both rotation and the magnetic field stabilize the fluid layer against convection. However, the marginal stability analysis of Chandrasekhar predicted that when rotation is present, adding a magnetic field destabilizes the fluid! A minimum of the critical Rayleigh number is obtained when the ratio of the Lorentz forces over the Coriolis forces (the Elsasser number) is of order one. This was confirmed by the pioneer experiments performed by Nakagawa (1957, 1958) with a layer of mercury placed inside a reconditioned cyclotron. The idea that the presence of the magnetic field could enhance convection when rotation is present has been rather influential in the past decades. It lead to the idea that convection

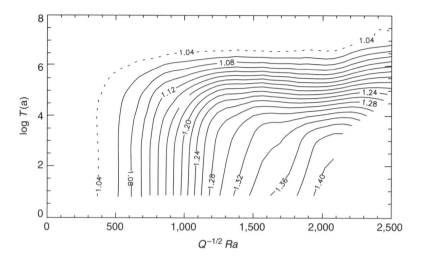

Figure 10 Isocontours of the Nusselt number for Rayleigh–Bénard convection in a horizontal layer of gallium in the presence of a vertical magnetic field **B** and rotation (with Ω vertical). $Q^{1/2}$ is proportional to B while $Ta^{1/2}$ is proportional to Ω. Note that both rotation and the magnetic field reduce the efficiency of heat transfer. There is no hint of a cooperative effect of the two in this regime. (Reprinted with permission from Aurnou, J. M. and Olson, P. L., "Experiments of Rayleigh–Bénard convection, magnetoconvection and rotating magneto-convection in liquid gallium," *J. Fluid Mech.* **430**, 283–307 (2001). Copyright (2001) by Cambridge University Press.)

could be more efficient in a rotating sphere if it creates a magnetic field, and hence that dynamo action was the natural expression of convection in a rotating sphere. Furthermore, it was argued that the amplitude of the magnetic field would settle at a level where the Coriolis and Lorentz forces are roughly in balance. These ideas are behind the proposal of a magnetostrophic experimental dynamo by our group in Grenoble (see above). Note that numerical models of convective dynamos only partly support this idea since the heat transfer is enhanced when a magnetic field is present (Christensen *et al.*, 1999; Grote *et al.*, 2000) but the onset of convection does not seem to be lowered (see Busse *et al.* in Chapter 6 of this volume). This discrepancy could be due to the fact that the magnetic field is not imposed in a dynamo, in contrast to the case treated by Chandrasekhar, or to the fact that numerical dynamos are not yet in the small Ekman number limit where his results apply. At any rate, it is clear that Rayleigh–Bénard convection with an imposed magnetic field and rotation is a topic that deserves further investigation. This motivated Aurnou and Olson (2001) to run a comprehensive set of experiments with liquid gallium. The focus was on measuring the Nusselt number (convective heat transfer divided by conductive: the Nusselt number is equal to unity in the absence of convection) for various values of the imposed magnetic field and rotation rate. As shown in Figure 10, Aurnou and Olson found that the measured Nusselt numbers could be simply contoured in the plane that draws a modified Rayleigh number along the x-axis and the logarithm of the Taylor number along the y-axis ($Ta = 4\Omega^2 D^4/\nu^2 = (2/E)^2$ where E is the Ekman number). The modified Rayleigh number $Ra\,Q^{-1/2}$ incorporates the effect of the magnetic field, since Q is the Chandrasekhar number $Q = \sigma B^2 D^2/\rho\nu$. The results indicate that convection is inhibited by the presence of rotation and by the magnetic field, as expected. In contrast with the experimental results of Nakagawa, there is no indication for a domain in which the simultaneous action of the Lorentz and Coriolis forces lead to an enhancement of convection. However, it should be noted that both rotation and the magnetic field are moderate in these new experiments. The asymptotic regime studied by Chandrasekhar and his followers is therefore not reached. A nice extension of the work of Aurnou and Olson would be to reach these asymptotic regimes and document both the heat transfer and velocity field. The results would be very valuable for understanding convective dynamos at small Prandtl number.

3.2. *Thermal convection in a rotating sphere and annulus*

The experimental study of thermal convection in a rotating sphere was pioneered by Busse and his coworkers (Carrigan and Busse, 1976). The centrifugal force can mimic the radial gravity of a self-gravitating planet. The sign is reversed, thus the temperature gradient must also be reversed. The fact that the centrifugal acceleration has a cylindrical symmetry is not important as long as one remains in the quasi-geostrophic regime, subject to the constraints of the Proudman–Taylor theorem. The asymptotic analyses of Roberts (1968) and Busse (1970) show that convection takes the form of convective columns with their axis aligned with the axis of rotation. The widths of the columns is proportional to $E^{1/3}$ where E is the Ekman number $E = \nu/\Omega R^2$ (ν is the kinematic viscosity, Ω is the rotation rate and R the radius of the sphere). The importance of the slope ζ of the spherical boundaries at the two ends of the columns was emphasized by Busse, who introduced a model of convection in an annulus with tilted top and bottom boundaries. The tilt is responsible for the development of thermal Rossby waves and bends the columns in a prograde direction. When the tilt varies with cylindrical radius as in the sphere, the period of the Rossby wave varies and produces a prograde spiralization of the columns (see Busse *et al.* in Chapter 6 of this volume). All these phenomena are well characterized in numerical models of convection just above the onset. Experiments have been used to extend the understanding of these phenomena farther away from the onset, and for small Ekman and Prandtl numbers.

3.3. *Thermal convection in an annulus filled with a low Prandtl number liquid*

Experiments were carried out by Lathrop and his colleagues with liquid sodium in a 20 cm diameter annulus spinning at rotation rates up to 6,000 rpm. By measuring the magnetic field induced in response to a sinusoidal excitation, Peffley *et al.* (2000b) found evidence for four columns at moderate Rayleigh numbers while a turbulent signature was observed at higher Rayleigh numbers. At these low Ekman numbers ($\sim 10^{-7}$), marginal stability theory (Jones *et al.*, 2000) predicts a much larger number of columns, which would probably not be detected in Peffley's experiments. It is not yet clear what causes the signal at lower frequency they observe.

Experiments with mercury and gallium–indium–tin alloy were run in Bayreuth, where the emphasis was on characterizing the onset of convection (Jaletzky, 1999). The liquid is contained in a small-gap cylindrical shell (height, $h = 50$ mm, gap width, $d = 5$ mm; middle radius in the gap, 67 mm) with upper and lower boundaries tilted at 45° ($\zeta = -1$). As predicted by marginal stability analysis, the onset is oscillatory. Rapid temperature oscillations are recorded on thermistors on the walls. However, the comparison of the measured frequencies and critical Rayleigh numbers with the theoretical predictions shows that one must take into account friction in the vertical Ekman boundary layers on the inner and outer cylinders

(Plaut and Busse, 2002). A suitable parameter is the Coriolis parameter $Co = (2\zeta d/h)/E$. For liquids with Prandtl number larger than about 10, dissipation on the vertical walls is negligible once the Coriolis parameter is larger than 10^4, but as the Prandtl number decreases one needs an increasingly large Coriolis parameter for dissipation on the walls to be negligible. In the weakly nonlinear regime for small Prandtl numbers, this theory also predicts a strong retrograde zonal wind, which remains to be searched for in the experiments.

3.4. Thermal convection in a rotating sphere filled with water and gallium

Experiments are well suited for exploring the fully developed convective states, which are difficult to access in numerical models. This route was followed by Cardin and Olson (1994) who studied convection in a sphere filled with water at high Rayleigh numbers. They found that convection becomes turbulent and that columns fill in a large part of the sphere, while their dimensions remain similar to those predicted at the onset. They observe a mean zonal flow, which is retrograde near the inner sphere and can be explained in terms of Reynolds stresses.

Further experiments have been carried out by our team in Grenoble (Aubert et al., 2001). Both water and liquid gallium were used to fill a 22 cm diameter sphere, which is spun up to 800 rpm. Ekman numbers range from 10^{-5} to 10^{-7}, while the Rayleigh number is up to 80 times critical. The Prandtl number Pr of water is 7 and that of gallium is 0.023 (see Table 1). The main objective was to derive scaling relationships in order to better identify physical mechanisms and permit an extrapolation to core conditions. Since gallium is opaque, optical measuring techniques are inappropriate. Therefore, we used ultrasonic Doppler velocimetry to measure convective velocities. Specific protocols were devised and validated for application to gallium (Brito et al., 2001). With this technique, radial profiles of radial velocity are obtained. We also measure the radial profile of average zonal velocity. The most striking observation is that a strong zonal retrograde flow is present in the gallium experiments. Following Cardin and Olson (1994), we attribute this to Reynolds stresses produced by the turbulent convective motions. Typical radial size δ_r and velocity u_r are deduced from the radial profiles of velocity and a local Reynolds number is built as $Re_l = u_r \delta_r / \nu$. While this number is less than 100 in all the water experiments, it reaches 600 in the gallium experiments, even though the Rayleigh number is at most four times critical in gallium. As in classical two-dimensional turbulence, the energy is not dissipated at the convective scale but cascades up to larger scales. In the sphere the only large-scale truly geostrophic motions are the zonal motions. These motions are created by the Reynolds stresses of the convective eddies and their velocity is limited by friction on the outer boundary. Following Cardin and Olson (1994), we developed a quasi-geostrophic turbulent model from which the following scaling relationships are derived:

$$u_r = \frac{\nu}{D^2} \left(\frac{Ra_Q}{Pr^2} \right)^{2/5} E^{1/5}, \tag{3}$$

$$\delta_r = D \left(\frac{Ra_Q}{Pr^2} \right)^{1/5} E^{3/5}, \tag{4}$$

$$\frac{u_{\text{zonal}}}{u_r} = Re_l^{2/3} E^{1/6}, \tag{5}$$

where we introduced the heat flux based Rayleigh number

$$Ra_Q = \frac{\alpha \gamma P D^2}{k \kappa \nu},$$

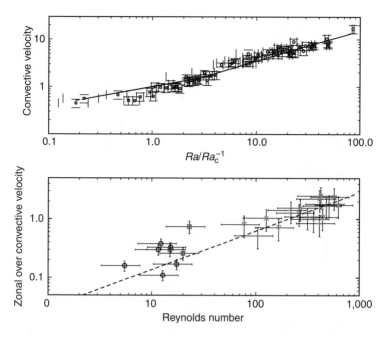

Figure 11 Thermal convection in a rotating sphere filled with water or gallium. Velocities are measured using ultrasonic Doppler velocimetry. Top: nondimensionalized radial velocity as a
function of Rayleigh number. All data points [squares for gallium (low *Ra*) and circles
for water (high *Ra*)] follow the law predicted from a quasi-geostrophic turbulent model
(solid line, see text). Bottom: ratio of the zonal to radial velocities increases with the local
Reynolds number (see text). Large zonal velocities are measured in the experiments for
gallium. (After Aubert *et al.*, 2001.)

in which P is the heat power input, D is the thickness of the spherical shell, α is the coefficient
of thermal expansion, k is the thermal conductivity, and γ is the acceleration (due to gravity
or to rotation). As shown in Figure 11, these predictions are found to be in excellent agreement with our measurements. These laws can be used to dimension a hypothetical convective
dynamo (see above). The large zonal velocities are reminiscent of the observations of Grote
et al. (2000) for convection in a rotating spherical shell with free-slip boundaries (see Busse
et al., Chapter 6). However, we do not observe the intermittency of convection that accompa-
nies this phenomenon in the numerical experiments. It could be because the mechanism that
limits the amplitude of the zonal flow is friction at the outer boundary in our experiments while
it is probably linked to viscous dissipation in the bulk of the liquid in the numerical experiments. Extrapolating to core conditions, we predict convective and zonal velocities of the
order of 1 mm s^{-1}. This is probably one order of magnitude larger than observed (Hulot *et al.*,
1990). Further experiments are under way to explore how an imposed magnetic field modifies
the physics and the laws (see Brito *et al.*, 1995). One of the interest of these experiments with
liquid metals is that flows are both turbulent (large Reynolds number) and strongly influenced
by rotation (small Ekman number). With measurement techniques such as ultrasonic Doppler
velocimetry, both the mean flow and the time dependence can be investigated quantitatively,
and we may hope to better characterize turbulence in this situation, and with a magnetic field as
well, one goal being to parameterize sub-grid turbulence in numerical models of the dynamo.

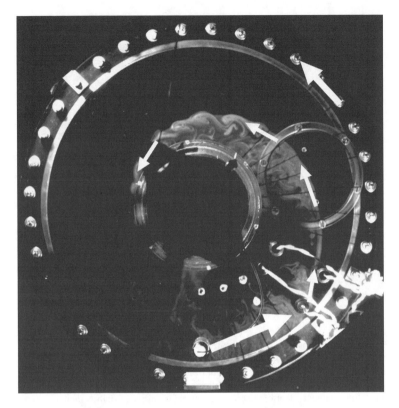

Figure 12 Equatorial planform of rotating hemispherical shell convection with anomalous heat flux at the outer boundary (heater, white rectangle) at $E = 4.7 \times 10^{-6}$, visualized by fluorescent dye. Rotation is anticlockwise. Global locking is observed here for $Ra/Ra_c = 26$ and a peak heat flux sixty-nine times larger than its mean. White dye was injected adjacent to the heater. White arrows indicate induced flows (eastward flow (broad arrow) and a narrow jet along the front (fine arrows)). (Reprinted with permission from Sumita, I. and Olson, P., "A laboratory model for convection in Earth's core driven by a thermally heterogeneous mantle," *Science* **286**, 1547–1549 (1999). Copyright (1999) American Association for the Advancement of Science.)

3.5. Core–mantle coupling

The question of thermal coupling between the core and the mantle has been studied experimentally by Sumita and Olson (1999). Because the time scales are so much larger in the mantle than in the core, and because the temperature fluctuations are very small in the core, the core–mantle boundary is seen as an isothermal boundary from the mantle side and as a boundary with an imposed stationary but spatially variable heat flux from the core side. This thermal inhomogeneity of the outer boundary can play a role on the organization of convective motions in the core. Sumita and Olson performed experiments in a hemisphere filled with water, rotating eastwards around a vertical axis. The thermal anomaly is provided by a narrow heater on the side. When the heat flux anomaly is strong (larger than about 35 times the mean), an interesting global locking is observed. As shown in Figure 12, a spiralling front forms, bordered by a narrow jet, which divides a warm region to the west from a cold region to the east, preventing mixing between them. The spiralling front originates eastwards of the heater, and a global eastward flow is observed as well. Velocities in the jet are four to five times larger than the eastward flow. Sumita and Olson suggest than fronts and jets could form in the core, in response to heat flux anomalies caused by mantle convection at the core–mantle boundary. These features could influence the dynamo and for example, control the path followed by the virtual paleomagnetic pole during inversions.

3.6. Crystallization experiments and the inner core

Rather unexpectedly, laboratory experiments have also been used to tackle the question of the solidification of the inner core. The goal is to explain the observation that seismic waves travelling through the inner core parallel to the axis of rotation of the Earth are some 4% faster than those that travel perpendicular to it (see Song, 1997 for a review). This is usually attributed to seismic anisotropy due to lattice preferred orientation of iron crystals. There remains to explain why the crystals are consistently aligned with respect to the axis of rotation and to account for the observed level of anisotropy.

The topic was investigated by Bergman (1997, 1998) in a series of papers. Bergman solidified tin–lead alloys under a controlled directional heat flow. He observed that dendrites tend to grow parallel to the heat flow, and correspond to a single crystallographic direction. This direction is $\langle 110 \rangle$ for the tin–lead alloy but is predicted to be $\langle 210 \rangle$ for iron in the hcp system believed to be favoured at inner core conditions. Bergman also measured ultrasonic wave anisotropy in the solidified alloy and found it to be in agreement with the solidification texturing. The extrapolation to the inner core assumes that heat flow is cylindrically radial so that crystals grow perpendicular to the axis of rotation. This is not incompatible with the organization of convection in vortices aligned with the axis of rotation, which transfer heat more efficiently parallel to the equatorial plane. Under this hypothesis, Bergman could reproduce the variation of seismic velocities as a function of the angle of propagation but found it difficult to account for a level of anisotropy as high as observed. From his experiments, Bergman also inferred that typical iron crystals in the inner core would be hundreds of meter wide and several kilometres long.

A slightly different approach was taken by Brito (1998). With Olson, he solidified a pure metal – gallium – under a directional heat flow and investigated the effect of several different ingredients that could play a role in the core. Using ultrasounds as well, they also found that crystals grow along the heat flow direction but not along a particular crystallographic axis. However, once a given crystallographic direction is present (from small pre-existing crystallites, for example) all crystals tend to grow with the same alignment. All other ingredients – stirring, magnetic field – could not change this behaviour. The extrapolation to the core suggests that a cylindrically radial heat flow is not sufficient to produce lattice preferred orientation in the inner core. Brito proposes that seismic anisotropy rather reflects some primordial crystallographic orientation.

4. Mantle convection

Experiments continue to play an important role in the study of mantle convection. In contrast to core dynamics, the proper nondimensional numbers can now be reached in three-dimensional numerical models of mantle convection. However, there are a number of complexities either asserted or presumed that are not dealt with easily in numerical models. They include strong variations of material properties, or at the opposite subtle chemical heterogeneities, both possibly leading to some kind of layering. Thermal plumes in a liquid with temperature-dependent viscosity also display specific features, such as oscillations akin to solitons in their conduit, which are suitable to analogical experiments.

4.1. Two-layer convection

In the 1970s, several studies were devoted to convection in two superposed layers (e.g. Richter and Johnson, 1974). At that time, there were clear indications that something was happening at a depth of about 650 km in the mantle, and it was thought that tectonic plates – the surface manifestation of mantle convection – could not subduct below that depth. Hence, the idea that mantle convection was organized in two superposed layers, the 650 km boundary marking some kind of compositional interface. One important issue was the kind of coupling that could exist between convective structures above and below the interface, and two end members were identified: 'mechanical coupling', when cold downwellings in the upper layer lie above hot uprisings in the lower layer, and 'thermal coupling' when hot uprisings in the two layers are superposed, yielding a large shear between cells at the interface (Cserepes *et al.*, 1988; Cardin *et al.*, 1991). Ten years later, it was clear that the 650 km seismic discontinuity was almost entirely due to a mineralogical phase transition rather than to compositional layering (although some questions have arisen recently (Hirose *et al.*, 2001)), and seismic tomography provided evidence that subducting slabs could penetrate the 650 km discontinuity and even reach the core–mantle boundary. However, the topic has been revived recently with emphasis on convection in layers that differ by a very small compositional difference. It is indeed plausible that the D″ layer or some part of the lower mantle has a slightly different chemical composition than the rest of the mantle. The density difference between the two layers would be very small and the interface would be widely distorted by convective motions. The situation is not easily dealt with in numerical models but can be investigated in the laboratory. A thorough investigation was carried out by Davaille (1999a) who superposed two miscible viscous liquids and varied the ratios between the densities, the viscosities, the depths and the vigour of convection. She observed several different regimes, depending mostly on the density ratio. Figure 13 displays two experiments with a strong density ratio (top) and a small density ratio (bottom). In the first case, the interface between the two layers is rather flat on average, but thin filaments of the opposite fluid are entrained on both sides. When the density ratio is weak, the interface is widely distorted and layers form diapirs that can extend all the way to the opposite boundary. In addition, the system can evolve with time from the first to the second situation since the density ratio between the two layers decreases as mixing proceeds. Davaille (1999b) noted that the features she observed are reminiscent of some geophysical observations: the thin filaments entrained by thermal plumes could explain peculiar geochemical signatures of hotspot basalts, while it is tempting to relate the large diapirs of the second regime with the superswells observed in seismic tomography. One important issue that remains to be further analysed is the role of chemical diffusion in these experiments. The inter-diffusivity of the two liquids is many orders of magnitude larger than chemical diffusivities in the mantle, and it is not completely clear yet how the entrainment process should be scaled. In turn, the experiments suggest that the boundary conditions usually employed in numerical models of layered convection probably oversimplify the real situation.

(a)

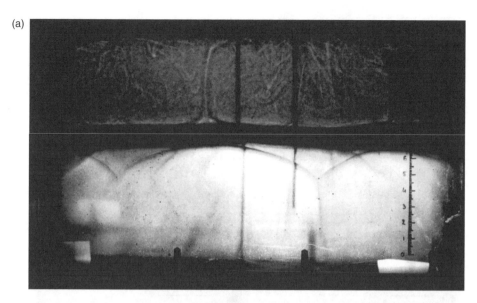

(b)

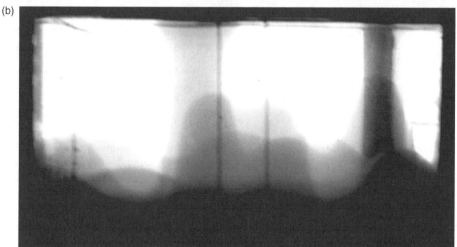

Figure 13 Patterns and entrainment in thermal convection of two superposed miscible liquids. Top: the density ratio of the two liquids is large compared to that of temperature induced variations. Thin filaments are advected away from convergence wedges at the interface. Bottom: the intrinsic density ratio is similar to that of temperature induced variations. Large domes form and can reach the opposite boundary before relaxing back. (Reprinted with permission from Davaille, A., "Simultaneous generation of hotspots and superswells by convection in a heterogenous planetary mantle," *Nature* **402**, 756–760 (1999). Copyright (1999) by *Nature*.) (See Colour Plate XIII.)

4.2. Plumes

Thermal plumes rising in the mantle could have a much lower viscosity than the surrounding mantle. Early experiments (e.g. Whitehead and Luther, 1975; Olson and Singer, 1985;

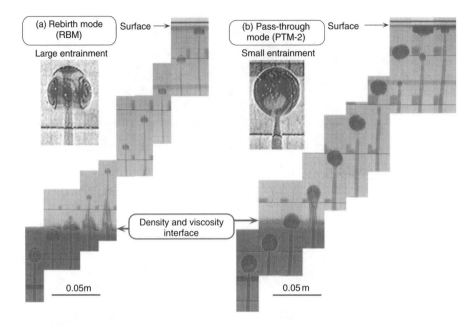

Figure 14 Composite images of the ascent history of a light plume across the interface between two miscible liquids. Two different behaviours are observed depending on the density and viscosity ratios: the rebirth mode (on the left) and the pass-through mode (on the right). (Reprinted from Kumagai, I. and Kurita, K., "On the fate of mantle plumes at density interfaces," *Earth Planet. Sci. Lett.*, **179**, 63–71, Copyright (2000) with permission from Elsevier Science.) (See Colour Plate XII.)

Griffiths and Campbell, 1990) showed that under these conditions, plumes tend to form a large head that rises at almost constant velocity and is fed by a narrow conduit. Recent experiments have dealt with the interaction of plumes with a density (and viscosity) interface. Figure 14 illustrates two different regimes discovered by Kumagai and Kurita (2000). When the density ratio between the ascending plume head and the upper layer is small, the plume is slowed down at the interface and a new plume forms and rises from the stagnating primary plume head. For larger density ratios, the plume is able to pass through the interface with little distortion. In the first mode, the ascent is slow and entrainment of lower layer material is strong in the plume head.

   Laudenbach and Christensen (2001) studied another aspect of plume dynamics. They injected hot glucose syrup (whose viscosity depends strongly on temperature) in cold syrup, and looked at the effect of a short enhancement of the injection rate. As illustrated in Figure 15, this produces a solitary conduit wave that propagates along the pre-existing plume conduit. By measuring the deflection of a laser beam, radial temperature profiles in the conduit and in the solitary wave can be obtained. They are found to be in good agreement with numerical models. Laudenbach and Christensen show that these waves have a closed flow structure (the material remains trapped within the wave) similar to that found in compositional plumes. The thermal wave propagates faster than 'normal' material in the plume conduit. It is also hotter and carries more material. Thermal solitary waves could therefore be at the origin of periodic magma production at weak hotspots.

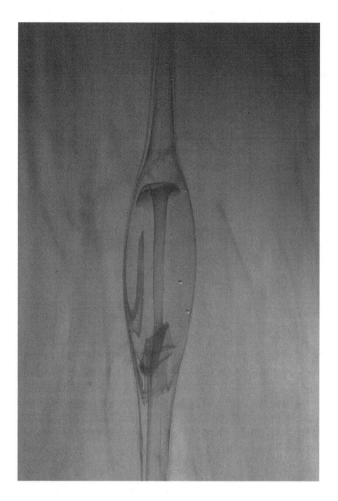

Figure 15 A solitary wave propagating in the conduit of a thermal plume. A sudden increase in the influx rate of a mature plume triggers the formation of a buldge that propagates upward in a solitary wave fashion. Dye is injected in the plume for visualization. (Reprinted with permission from Laudenbach and Christensen, 2001). (See Colour Plate XIV.)

4.3. Convection in a radial force field

Every experimentalist has once dreamed of reproducing in the laboratory the radial gravity field of self-gravitating bodies such as the Earth in order to investigate the effect of spherical geometry on convection. One way of approaching this is to use a hemisphere and spin it at a rotation rate just enough for the centrifugal acceleration at the equator to match the acceleration due to gravity. However, one is limited to a hemisphere and the acceleration equipotentials are not spherical. Nevertheless, this has proven useful in the case of experiments where convection is strongly influenced by rotation (small Ekman number) (Cordero and Busse, 1992; Cordero, 1993; Sumita and Olson, 1999, 2000).

Two other tracks have been recently followed. In both cases, gravity is replaced by another body force.

A group in Bremen is building a spherical shell in which the dielectrophoretic force is driving convection (Egbers *et al.*, 1999). The temperature difference is replaced by a difference in electrical potential between the inner and outer sphere. The experiment is meant to function in space in microgravity, as did the pioneer experiment of Hart *et al.* (1986). The platform of convection will be observed through a clever shadowgraph technique.

An even more far-fetched idea was developed by Rosensweig *et al.* (1999). Specialists of ferrofluids, the researchers of that group filled a spherical shell with ferrofluid and placed a magnet within the inner sphere, surrounded by a heating foil. Under these conditions, convection can take place. The fact that ferrofluids are opaque makes the observation of planforms more difficult. The faint temperature fluctuations caused by convection at the surface were observed using an infrared camera.

Although one is amazed by the deployment of imagination demonstrated in these experiments, it is not clear that they will help understanding better the organization of convection in planetary mantles. Indeed, the central acceleration is very different from that of planets: it decays as r^{-5} for the dielectrophoretic force and r^{-4} for the ferrofluid set-up (where r is radius), while it is constant in the Earth's mantle and increases almost linearly with radius in the core.

5. Acknowledgements

I thank G. Gerbeth, U. Müller, R. Stieglitz, D. Lathrop, J. Aurnou, J. Aubert, I. Sumita, A. Davaille, K. Kurita and N. Laudenbach for providing the figures reprinted in this review.

References

Aubert, J., Brito, D., Nataf, H.-C., Cardin, P. and Masson, J.-P., "Scaling relationships for finite amplitude convection in a rapidly rotating sphere, from experiments with water and gallium," *Phys. Earth Planet. Inter.* **128**, 51–74 (2001).

Aurnou, J. M. and Olson, P. L., "Experiments on Rayleigh–Bénard convection, magnetoconvection and rotating magnetoconvection in liquid gallium," *J. Fluid Mech.* **430**, 283–307 (2001).

Bergman, M. I., "Measurements of elastic anisotropy due to solidification texturing and the implications for the Earth's inner core," *Nature* **389**, 60–63 (1997).

Bergman, M. I., "Estimates of the Earth's inner core grain size," *Geophys. Res. Lett.* **25**, 1593–1596 (1998).

Bourgoin, M., Marié, L., Petrelis, F., Burguete, J., Chiffaudel, A., Daviaud, F., Fauve, S., Odier, P., Pinton, J.-F., "MHD measurements in the von Kármán sodium experiment," *Phys. Fluids* **14** (9), 3046 (2001).

Brito, D., "Approches expérimentales et théoriques de la dynamique du noyau terrestre," *Thèse Université Paris* 7, Janvier 1998.

Brito, D., Cardin, P., Nataf, H.-C. and Marolleau, G., "Experimental study of a geostrophic vortex of gallium in a transverse magnetic field," *Phys. Earth Planet. Inter.* **91**, 77–98 (1995).

Brito, D., Nataf, H.-C., Cardin, P., Aubert, J. and Masson, J.-P., "Ultrasonic Doppler velocimetry in liquid gallium," *Exp. Fluids* **31**, 653–663 (2001).

Bullard, E. C. and Gellman, H., "Homogeneous dynamos and terrestrial magnetism," *Phil. Trans. R. Soc. Lond.* A **247**, 213–278 (1954).

Busse, F. H., "Thermal instabilities in rapidly rotating systems," *J. Fluid Mech.* **44**, 441–460 (1970).

Busse, F. H., Müller, U., Stieglitz, R. and Tilgner, A., "Spontaneous generation of magnetic fields in the laboratory," In: *Evolution of Spontaneous Structures in Dissipative Continuous Systems* (Eds F. H. Busse and S. C. Müller), pp. 546–558. Springer, New York (1998a).

Busse, F. H., Grote, E. and Tilgner, A., "On convection driven dynamos in rotating spherical shells," *Stud. Geophys. Geod.* **42**, 1–6 (1998b).

Cardin, P. and Olson, P., "Chaotic thermal convection in a rapidly rotating spherical shell: consequences for flow in the outer core," *Phys. Earth Planet. Inter.* **82**, 235–239 (1994).

Cardin, P., Nataf, H.-C. and Dewost, P., "Thermal coupling in layered convection: evidence for an interface viscosity control from mechanical experiments and marginal stability analysis," *J. Phys. II* **1**, 599–622 (1991).

Cardin, P., Brito, D., Jault, D., Nataf, H.-C. and Masson, J.-P., "Towards a rapidly rotating liquid sodium dynamo experiment," *Magnetohydrodynamics* **38**, 177–189 (2002).

Carrigan, C. R. and Busse, F. H., "Laboratory simulation of thermal convection in rotating planets and stars," *Science* **191**, 81–83 (1976).

Chandrasekhar, S., *Hydrodynamic and Hydromagnetic Stability*. Oxford University Press, Oxford (1961).

Christensen, U., Olson, P. and Glatzmaier, G. A., "Numerical modelling of the geodynamo: a systematic parameter study," *Geophys. J. Int.* **138**, 393–409 (1999).

Colgate, S. A., Li, H. and Pariev, V., "The origin of the magnetic fields of the universe: the plasma astrophysics of the free energy of the universe," *Phys. Plamas* **8**, 2425–2431 (2001a).

Colgate, S. A., Beckley, H. F., Pariev, V. I., Finn, J. M., Li, H., Weatherall, J. C., Romero, V. D., Westpfahl, D. J. and Ferrel, R., "The New Mexico liquid sodium α–Ω experiment," http://kestrel.nmt.edu/ dynamo/ (2001b).

Cordero, S., "Experiments on convection in a rotating hemispherical shell: transition to chaos," *Geophys. Res. Lett.* **20**, 2587–2590 (1993).

Cordero, S. and Busse, F. H., "Experiments on convection in rotating hemispherical shells: transition to a quasi-periodic state," *Geophys. Res. Lett.* **19**, 733–736 (1992).

Cowling, T. G., "The magnetic field of sunspots," *Mon. Not. R. Astr. Soc.* **94**, 39–48 (1934).

Cserepes, L., Rabinowicz, M. and Rosemberg-Borot, C., "3-dimensional infinite Prandtl number convection in one and 2 layers with implications for the Earth gravity-field," *J. Geophys. Res.* **93**, 12009–12025 (1988).

Davaille, A., "Two-layer thermal convection in miscible viscous fluids," *J. Fluid Mech.* **379**, 223–253 (1999a).

Davaille, A., "Simultaneous generation of hotspots and superswells by convection in a heterogenous planetary mantle," *Nature* **402**, 756–760 (1999b).

Dudley, M. L. and James, R. W., "Time-dependent kinematic dynamos with stationary flows," *Proc. R. Soc. Lond.* A **425**, 407–429 (1989).

Egbers, C., Brasch, W., Sitte, B., Immohr, J. and Schmidt, J. R., "Estimates on diagnostic methods for investigations of thermal convection between spherical shells in space," *Meas. Sci. Technol.* **10**, 866–877 (1999).

Elsasser, W. M., "Induction effects in terrestrial magnetism. 1. Theory," *Phys. Rev.* **69**, 106–116 (1946).

Fearn, D. R., "The geodynamo," In: *Earth's Deep Interior* (Ed. D. J. Crossley), *The Fluid Mechanics of Astrophysics and Geophysics*, **7**, pp. 79–114. Gordon and Breach, New York (1998).

Forest, C., "Madison dynamo experiment," http://aida.physics.wisc.edu/ (1998).

Frick, P., Denisov, S., Khripchenko, S., Noskov, V., Sokoloff, D., Stepanov, R. and Sukhanovsky, A., "A nonstationary dynamo experiment (current state of Perm project)," In: *Proceedings of the Fourth International Pamir Conference*, Presqu'île de Giens, pp. 183–188. France (2000).

Gailitis, A. and Freibergs, J., "Theory of a helical MHD dynamo," *Magnetohydrodynamics* **12**, 127–129 (1976).

Gailitis, A., Karasev, B. G., Kirillov, I. R., Lielausis, O. A., Luzhanskii, S. M., Ogorodnikov, A. P. and Preslitskii, G. V., "Liquid metal MHD dynamo model experiment," *Magnetohydrodynamics* **23**, 349–353 (1987).

Gailitis, A., Lielausis, O., Dement'ev, S., Platacis, E., Cifersons, A., Gerbeth, G., Gundrum, T., Stefani, F., Christen, M., Hänel, H. and Will, G., "Detection of a flow induced magnetic field eigenmode in the Riga dynamo facility," *Phys. Rev. Lett.* **84**, 4365–4368 (2000).

Gailitis, A., Lielausis, O., Platacis, E., Dement'ev, S., Cifersons, A., Gerbeth, G., Gundrum, T., Stefani, F., Christen, M. and Will, G., "Magnetic field saturation in the Riga dynamo experiment," *Phys. Rev. Lett.* **86**, 3024–3027 (2001).

Glatzmaier, G. A., "Numerical simulations of a convective dynamo experiment," In: *SEDI 2000*, The 7th symposium of Study of the Earth's Deep Interior, Exeter, UK, Abstract S5.7 (2000).

Griffiths, R. W. and Campbell, I. H., "Stirring and structure in mantle starting plumes," *Earth Planet. Sci. Lett.* **99**, 66–78 (1990).

Grote, E., Busse, F. H. and Tilgner, A., "Regular and chaotic spherical dynamos," *Phys. Earth Planet. Inter.* **117**, 259–272 (2000).

Hart, J. E., Glatzmaier, G. A. and Toomre, J., "Space-laboratory and numerical simulations of thermal convection in a rotating hemispherical shell with radial gravity," *J. Fluid Mech.* **173**, 519–544 (1986).

Herzenberg, A., "Geomagnetic dynamos," *Phil. Trans. R. Soc. Lond.* A **250**, 543–585 (1958).

Hirose, K., Fei, Y., Ono, S., Yagi, T. and Funakoshi, K.-I., "In situ measurements of the phase transition boundary in $Mg_3Al_2Si_3O_{12}$: implications for the nature of the seismic discontinuities in the Earth's mantle," *Earth Planet. Sci. Lett.* **184**, 567–573 (2001).

Hulot, G., Le Mouël, J.-L. and Jault, D., "The flow at the core–mantle boundary: symmetry properties," *J. Geomagn. Geoelectr.* **42**, 857–874 (1990).

Jaletzky, M. U., "Über die Stabilität von thermisch getriebenen Strömungen im rotierenden konzentrischen Ringspalt," Doktorat Universität Bayreuth (1999).

Jault, D., Nataf, H.-C. and Cardin, P., "Feasability study of a dynamo experiment in the magnetostrophic regime," In: *SEDI 98*, the 6th Symposium of Study of the Earth's Deep Interior, Tours, France, Abstract S7.9 (1998).

Jones, C. A., Soward, A. M. and Mussa, A. I., "The onset of thermal convection in a rapidly rotating sphere," *J. Fluid Mech.* **405**, 157–179 (2000).

Kumagai, I. and Kurita, K., "On the fate of mantle plumes at density interfaces," *Earth Planet. Sci. Lett.* **179**, 63–71 (2000).

Larmor, J., "How could a rapidly rotating body such as the Sun become a magnet?" *Brit. Assn. Adv. Sci. Rep.* **1919**, 159–160 (1919).

Laudenbach, N. and Christensen, U. R., "An optical method for measuring temperature in laboratory models of mantle plumes," *Geophys. J. Int.* **145**, 528–534 (2001).

Lielausis, O., "Dynamo theory and liquid-metal MHD experiments," *Astron. Nachr.* **315**, 303–317 (1994).

Lowes, F. J. and Wilkinson, I., "Geomagnetic dynamo: a laboratory model," *Nature* **198**, 1158–1160 (1963).

Lowes, F. J. and Wilkinson, I., "Geomagnetic dynamo: an improved laboratory model," *Nature* **219**, 717–718 (1968).

Nakagawa, Y., "Experiments on the instability of a layer of mercury heated from below and subject to the simultaneous action of a magnetic field and rotation," *Proc. R. Soc. Lond.* A **242**, 81–88 (1957).

Nakagawa, Y., "Experiments on the instability of a layer of mercury heated from below and subject to the simultaneous action of a magnetic field and rotation II," *Proc. R. Soc. Lond.* A **249**, 138–145 (1958).

Olson, P. and Singer, H., "Creeping plumes," *J. Fluid Mech.* **158**, 511–531 (1985).

Peffley, N. L., Cawthorne, A. B. and Lathrop, D. P., "Toward a self-generating magnetic dynamo: the role of turbulence," *Phys. Rev. E* **61**, 5287–5294 (2000a).

Peffley, N. L., Goumilevski, A. G., Cawthorne, A. B. and Lathrop, D. P., "Characterization of experimental dynamos," *Geophys. J. Int.* **142**, 52–58 (2000b).

Plaut, E. and Busse, F. H., "Low-Prandtl-number convection in a rotating cylindrical annulus," *J. Fluid Mech.* **464**, 345–363 (2002).

Ponomarenko, Y. B., "Theory of the hydromagnetic generator," *J. Appl. Mech. Tech. Phys.* **14**, 775–778 (1973).

Rädler, K.-H., Apstein, E., Rheinhardt, M. and Schüler, M., "The Karlsruhe dynamo experiment. A mean field approach," *Stud. Geophys. Geod.* **42**, 224–231 (1998).

Richter, F. M. and Johnson, J. E., "Stability of a chemically layered mantle," *J. Geophys. Res.* **79**, 1635–1639 (1974).

Roberts, G. O., "Dynamo action of fluid motions with two-dimensional periodicities," *Phil. Trans. R. Soc. Lond.* A **271**, 411–454 (1972).

Roberts, P. H., "On the thermal instability of a rotating fluid sphere containing heat sources," *Phil. Trans. R. Soc. Lond.* A **263**, 93–117 (1968).

Roberts, P. H. and Glatzmaier, G. A., "Geodynamo theory and simulations," *Rev. Mod. Phys.* **72**, 1081–1123 (2000).

Rosensweig, R. E., Browaeys, J., Bacri, J.-C., Zebib, A. and Perzynski, R., "Laboratory study of spherical convection in simulated central gravity," *Phys. Rev. Lett.* **83**, 4904–4907 (1999).

Song, X., "Anisotropy of the Earth's inner core," *Rev. Geophys.* **35**, 297–314 (1997).

Stieglitz, R. and Müller, U., "Experimental demonstration of a homogeneous two-scale dynamo," *Phys. Fluids* **13**, 561–564 (2001).

Sumita, I. and Olson, P., "A laboratory model for convection in Earth's core driven by a thermally heterogeneous mantle," *Science* **286**, 1547–1549 (1999).

Sumita, I. and Olson, P. L., "Laboratory experiments on high Rayleigh number thermal convection in a rapidly rotating hemispherical shell," *Phys. Earth Planet. Inter.* **117**, 153–170 (2000).

Sweet, D., Ott, E., Antonsen, Jr, T. M. and Lathrop, D. P., "Blowout bifurcations and the onset of magnetic dynamo action," *Phys. Plasmas* **8**, 1944–1952 (2001).

Tilgner, A., "A kinematic dynamo with a small scale velocity field," *Phys. Lett.* A **226**, 75–79 (1997).

Tilgner, A., "Towards experimental fluid dynamos," *Phys. Earth Planet. Inter.* **117**, 171–177 (2000).

Tilgner, A. and Busse, F. H., "Saturation mechanism in a model of the Karlsruhe dynamo," In: *Dynamo and Dynamics, a Mathematical Challenge* (Eds P. Chossat, D. Armbouster and I. Oprea) NATO Sci. Ser. II Mathematics, Physics and Chemistry **26**, pp. 153–161. Kluwer Academic Publishers, Dordrecht (2001).

Whitehead, J. A. and Luther, D. S., "Dynamics of laboratory diapir and plume models," *J. Geophys. Res.* **80**, 705–717 (1975).

8 Dynamics of the core at short periods

Theory, experiments and observations

Keith D. Aldridge[1]

Oxford Centre for Industrial and Applied Mathematics, University of Oxford, OX1 3LB, UK

Traditional linearized theory of contained, rotating, homogeneous fluids ignored phenomena with Rossby number, $\varepsilon \ll 1$, for $E \to 0$ where $E = \nu/\Omega L^2$ is the Ekman number defined in terms of ν, the fluid's kinematic viscosity, Ω, its rotation speed and L, its length scale. We are reminded of the error produced by this assumption when we see the onset of a parametric instability, even when the Rossby number, $\varepsilon \approx E^{1/2}$. This inertial instability provides a mechanism to produce very long time-scale response from a short-period perturbation. If one ignores the possibility of an instability, a class of possible explanations of an observed phenomena will be overlooked: this is precisely why some of the early experiments in precession were incompletely interpreted and part of the basis for rejecting precession as a way to maintain the geodynamo.

Revival of interest in inertial phenomena has been fueled significantly by attention from the mathematical community, with its interest dating back to ill-posed boundary value problems, their associated discontinuous solutions and properties of characteristic surfaces. While it might appear to be only of mathematical interest, the importance of these phenomena at the low Ekman numbers likely to be found in the Earth's fluid core, means that the physical counterpart of the analytical solutions may play an important role in core dynamics. Furthermore, the possibility of inertial and subsequent secondary instabilities in the Earth's core is a relatively new subject with its possible significance for the geodynamo still being studied. In this regard, it has been suggested that structural similarity between the steady flow observed in experiments on precession of barotropic fluids and that of columnar convection, could couple the two processes. We also consider differential rotation of the Earth's solid inner core, likely carrying with it a tangent cylinder, a phenomena that has been studied theoretically and observed experimentally. Recent comparison of the polar geomagnetic field at two epochs has confirmed the likely existence of such an anticyclonic vortex in the fluid outer core, consistent with results from numerical models of the geodynamo.

While inference of core motion has been possible through observations of changes in the Earth's geomagnetic field, such as the one described above, another window has been opened on both the inner and outer core through the use of very long baseline interferometry (VLBI) and superconducting gravimeters (SGs). Both of these tools have already proven their precise capabilities in measurement of Earth wobble and nutational frequencies. Oscillations of the solid inner core, the Slichter modes, while once considered too small to be observed, are now at the threshold of detection through the use of SG. Identification of short-period, large-scale

---

[1] Permanent address: Centre for Research in Earth and Space Science, York University, Toronto, Canada, M3J 1P3.

fluid oscillations of the outer core could be possible through the combined use of VLBI and SG with extensive data sets currently being collected through the Global Geodynamics Project (GGP).

The theme of this work, as outlined above, will be developed around the many contributions to theory, experiments and observations related to dynamics of the Earth's core.

1. Introduction

Geophysical interest in the Earth's fluid outer core is motivated mainly by the problem of the origin and maintenance of the geomagnetic field produced there. While oscillations (Smylie *et al.*, 1993, 2001) and possible super-rotation of the essentially solid inner core (Gubbins, 1981; Song and Richards, 1996; Su *et al.*, 1996; Creager, 1997; Souriau *et al.*, 1997; Souriau, 1998) have occupied several recent studies of the the deep interior of the Earth, it is the existence of an inner boundary to the fluid core which has attracted attention of both the geophysical and mathematical communities since for the simplest case of a homogeneous fluid, there may be no continuous inertial mode solutions (Rieutord, 1995; Rieutord and Valdettaro, 1997). The dynamical behaviour of a conducting fluid of small viscosity, contained by a rapidly rotating, precessing mantle in a spheroidal shell is indeed, a multi-faceted problem. Its solution has followed several approaches including both experiments and theory in combination with observations. This paper centres on these three aspects related to rotation of homogeneous fluids while the baroclinic and hydromagnetic parts of the problem are considered by Nataf in Chapter 7 of this volume.

A contained, rotating, constant density fluid supports a spectrum of inertial modes which are travelling waves when the azimuthal wavenumber is greater than or equal to one (Greenspan, 1968). The frequencies of these modes lie between zero and twice the rotation speed, the latter case corresponding to particle motions which are predominantly in planes perpendicular to the rotation axis, while near zero frequency particle motions are essentially parallel to the axis of rotation. If the fluid is self-gravitating and slightly stratified so that the Brunt–Väisälä frequency, N, is close to zero, the modes are gravity-inertial waves with frequencies modified to fit in a range which is now bounded below by N rather than zero. While the value of N is unknown for the Earth's core it is expected to be small but likely variable with radius. In this case the effects of rotation will be dominant with the coriolis force greater than the buoyancy force so that important properties of core dynamics can be appreciated by understanding rotational effects. These are reviewed below with special reference to dissipation.

Although most dissipation of inertial modes excited through perturbation of container boundaries takes place in boundary layers, large amplitudes can be produced in the fluid interior. Experiments by Aldridge and Toomre (1969) confirmed this fact through observations of low order spatial modes and their theoretical interpretation. This result is important for understanding of dynamics of the Earth's core because it had been concluded by Loper (1975) that mantle precession was insignificant in its effects on the deep core motions since almost all the energy was dissipated in boundary layer regions near the core–mantle boundary (CMB). While it is correct that energy is indeed dissipated there, it is wrong to assume that this implies no significant velocity field in the bulk of the core. Section 2.2 gives the details of this argument.

Also ignored in these earlier studies of the effects of boundary driven flows in the Earth's core, is the possibility of a parametric instability of tidal origin. For example, the precession of the mantle at a period of 25,800 years produces a shear in the fluid interior which will indeed be unstable if not prevented by viscous or Lorentz dissipation (Kerswell, 1993). Additionally, the semi-diurnal tide can produce a parametric elliptical instability, depending on dissipation effects, on time scales as short as a few thousand years.

Laminar flow can give way to turbulence in either of the above cases as demonstrated by experiments in both precession (Malkus, 1968; Vanyo *et al.*, 1995; Noir, 2000) and semi-diurnal tide-like forcing (Aldridge *et al.*, 1997; Seyed-Mahmoud, 1999; Seyed-Mahmoud *et al.*, 2000) as predicted from theoretical work on these instabilities by Kerswell (1993). It has been shown, and verified experimentally, that the growth rate of the instability scales with the amplitude of the boundary disturbance and the rotation speed of the container. Thus, the instability mechanism can produce a fluid response on a geomagnetic time-scale from a perturbation on a diurnal time scale.

While experimental observations support the existence of these parametric instabilities, it appears that they can be overwhelmed by secondary instabilities. Recent work by Eloy *et al.* (2000) has produced what appear to be finite amplitude standing waves and these are what are referred to as saturated states. In this case, an instability does not lead to a turbulent response but rather the flow reaches a large amplitude while remaining laminar.

We are led to speculate that for sufficiently low dissipation rates, perturbation of fluid planetary cores by tidal forces could initiate a parametric instability, as has been argued by Kerswell and Malkus (1998) for their explanation of the Io's magnetic signature, while it has been argued by Malkus (1994) that inertial instability is likely to be important for the outer planets. In the experiments designed to excite the instability, however, large amplitude geostrophic flows associated with secondary instability are seen to develop. It seems highly possible that these flows could be coupled to columnar convective flow thought to be integral to powering the geodynamo, so that the instability of rotational origin could in this case couple with the instability of gravitational origin. Thus rotational energy released through a parametric or secondary instability could become a control on a convective instability in maintaining the geomagnetic dynamo.

2. Theoretical background

2.1. Governing equations

We consider first the simplest case of a contained, homogeneous incompressible fluid in steady rotation about a fixed axis. The Navier–Stokes equation for the velocity **u** and reduced pressure p of a fluid of density ρ and kinematic viscosity ν referred to the rotating frame of reference $\Omega\mathbf{k}$ expresses the conservation of momentum as

$$\frac{\partial \mathbf{u}}{\partial t} + (\mathbf{u} \cdot \nabla)\mathbf{u} + 2\Omega\mathbf{k} \times \mathbf{u} = -\frac{1}{\rho}\nabla p + \nu\nabla^2\mathbf{u}$$

and conservation of mass is given by

$$\nabla \cdot \mathbf{u} = 0.$$

By defining the Rossby number

$$\varepsilon = U/\Omega L,$$

the Ekman number

$$E = \nu/\Omega L^2$$

and scaling length, time, velocity and pressure with L, Ω^{-1}, U and $\varepsilon\rho\Omega^2 L^2$, respectively the above momentum and continuity equations in the scaled variables are

$$\frac{\partial \mathbf{u}}{\partial t} + \varepsilon(\mathbf{u} \cdot \nabla)\mathbf{u} + 2\mathbf{k} \times \mathbf{u} = -\nabla p + E\nabla^2\mathbf{u}$$

and

$$\nabla \cdot \mathbf{u} = 0,$$

which are solved subject to the condition that

$$\mathbf{u} = \mathbf{0}$$

on the solid boundaries.

2.2. Modal excitation and dissipation

We review the calculation of the amplitudes of inertial modes. To fix ideas we consider the axially symmetric modes of oscillation in a rotating spherical cavity, excited by adding a periodic component ω to the steady rotation:

$$\Omega(t) = \Omega + \varepsilon\omega\cos\omega t,$$

as described by Aldridge and Toomre (1969).

Fluid is alternately pumped into and withdrawn from the interior as the container alternately slows down and speeds up at frequency ω. If the frequency ω is chosen to coincide with one of the eigenfrequencies

$$\omega_{nmk} = \lambda_{nmk}\Omega,$$

where λ_{nmk} is the mth root of

$$k P_n^k \left(\frac{\lambda}{2}\right) = 2\left(1 - \frac{\lambda^2}{4}\right)\frac{dP_n^k}{d\lambda}\left(\frac{\lambda}{2}\right),$$

calculated from the above boundary value problem and the boundary is well coupled to the velocity field of the eigenmode p_{nmk}, a resonance of the fluid will be established. The amplitude of the fluid's response at resonance is determined by the condition that the rate at which work is being done on the fluid must balance the rate at which energy is dissipated in the boundary layer. At first, we neglect the energy dissipated in the fluid interior and correct for that later. We find that this balance determines the amplitudes M, N of the modes as

$$(2I_2 M + I_1)^2 + (2I_2 N + I_{1s})^2 = I_1^2 + I_{1s}^2,$$

where I_q with $q = 1, 1s, 2$ are integrals over functions of the interior velocity field of the mode (Aldridge and Toomre, 1969).

A formal method to determine the dissipation of inertial modes in a sphere can be found in Greenspan (1968). Here we give a more physical description of the dissipation mechanism and how to obtain the modal decay rates. In the above experiment, the decay of the modes can be found from energy considerations: the e-folding time t_{nmk} of the mode with eigenfrequency ω_{nmk} will be twice the kinetic energy of the mode divided by the rate of dissipation in the boundary layer. This works out as

$$t_{nmk} = E^{-1/2}(I_e/I_2)\omega_{nmk}^{-1},$$

where I_e is an integral which determines the kinetic energy of the mode.

Table 1 Dissipation integrals for
some low-order modes

(m, n)	I_2	I_d
$(1, 1)$	2.203	31.3
$(2, 1)$	1.453	47.3
$(3, 1)$	1.262	71.9
$(4, 1)$	1.184	103.8

The amplitude of an inertial wave excited via a boundary is determined by the balance between work done by the container and dissipation in the boundary layer. Even though most of the dissipation takes place at the boundary, large amplitudes can exist in the fluid interior. In general, this means that

$$\frac{D_{int}}{D_{bl}} \approx O(E^{1/2}).$$

In the specific case of an axially symmetric, oscillatory perturbation $\varepsilon\omega\cos(\omega t)$ superimposed on a steady rotation Ω, for a sphere of radius a filled with fluid of kinematic viscosity ν and density ρ, these dissipations become

$$D_{int} = \rho\nu\varepsilon^2\omega_{mn}^2 a^3 (M^2 + N^2) I_d$$

and

$$D_{bl} = \rho\varepsilon^2\omega_{mn}^2 a^4 (\omega_{mn}\nu)^{1/2}(M^2 + N^2) I_2$$

so

$$\frac{D_{int}}{D_{bl}} = E^{1/2}\frac{I_d}{I_2}.$$

For a few low-order modes of frequency ω_{mn}, and amplitude M, N excited in a fluid-filled, spherical cavity, values of the dissipation integrals I_2, I_d are given in Table 1.

If the amplitude discrepancy between observation and linear theory in Figure 1 is taken to be 4% for the (1,1) mode, with $I_d/I_2 \cong 14$, $\varepsilon \to 0$, the other I_d/I_2 discrepancies suggest proportionate reductions of about 9, 16 and 25% for the (2,1), (3,1) and (4,1) response peaks, respectively. This trend can be seen in Figure 3 of Aldridge and Toomre (1969).

If one were to estimate the fluid's response in the laboratory experiment by energy considerations alone, it might be tempting to conclude that not much could be happening in the fluid interior because, as we have just seen, almost all the dissipation takes place in the very thin Ekman boundary layers. But in reality velocity amplitudes in the deep interior of the fluid are of the same order as the boundary excitation and the dissipation, while small in the interior, takes place almost entirely in the boundary layer. Based on energy considerations, it would be equally erroneous in a geophysical application to conclude that because most of the energy of a fluctuating mantle was dissipated near the CMB, that one could ignore the effects of those same mantle fluctuations on the deep Earth velocity field.

3. Parametric instabilities

Scaling of the equations of motion, while certainly a well-proven method for retaining only terms of relevant size for a primary flow, fails to retain terms which lead to a parametric instability. This happens because the size of second-order terms that account for an instability can be indeed much smaller than the linear terms, but their accumulated effect through continued energy input can easily overwhelm the first-order terms.

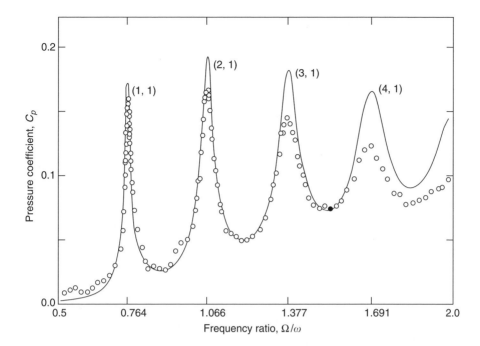

Figure 1 Response at resonance. (Reprinted with permission of the Cambridge University Press, from Aldridge, K. D. and Toomre, A. "Axisymmetric inertial oscillations of a fluid in a rotating spherical container," *J. Fluid Mech.* **37**, 307–323 (1969).)

Parametric instabilities have been observed in experiments on rotating fluids dating back to the observations of Malkus (1968) on precessing spheroids and theoretical work by Stewartson and Roberts (1963). At the time of Malkus' (1968) work, however, an interpretation by Busse (1968) based on a non-linear boundary layer analysis appeared to explain these observations. More recently, theoretical work by Kerswell (1993) offers an interpretation of Malkus' observations as a parametric instability. This theoretical work by Kerswell (1993) on what are termed inertial instabilities, provided a theoretical basis for understanding of instability observed in several previous experiments, as had been conjectured by Lumb *et al.* (1993). The basis of Kerswell's model is that pairs of inertial modes can be coupled through either an elliptical distortion of otherwise circular streamlines or through axial shearing of the streamlines. In either case, instability will occur as long as the rate of its growth exceeds its decay rate.

One way to classify the experiments in which instability has been observed is to group them according to the rotation and perturbation axes. We shall see that the instability can have mixed elliptical and shear distortion so that it is perhaps clearer to group the observations with respect to the the manner in which the instability was produced.

3.1. Coincident axes

In the first case we can consider experiments in which the two axes are coincident. In a set of experiments to study inertial waves, McEwan (1970) also observed what appears to have been a shear instability in a fluid in a steadily rotating right circular cylinder. Both the waves and instability were excited by perturbing the fluid cylinder with slightly tilted lid, coaxially with the rotation. McEwan (1970) had excited inertial modes using this perturbation, and accordingly he called the phenomenon 'resonant collapse'. An example of this collapse is shown in time sequence of photographs from McEwan (1970) shown here in Figure 2. The fluid in the cylindrical cavity contains reflective particles and shows the characteristic development of an 'S' shape, stationary in the perturbation (laboratory) frame from top left to bottom right as time passes. Collapse has been reached in the last panel.

A simpler version of the above experiment was carried out by Thompson (1970) who rotated a right circular cylinder and then tilted it slightly to effectively produce a 'diurnal tide'. This undulation of the free surface produced the shear instability, as evidenced by unstable or wavy disturbances at very small tilts for certain critical heights. Shown in Figure 3 is Thompson's (1970) Figure 4, illustrating the occurrence of instability by solid squares in the figure.

In a similar experiment designed primarily to measure inertial wave frequencies of a fluid cylinder during spin-up from rest, Stergiopoulos and Aldridge (1987) also observed this shear instability near resonance both during spin-up and in the state of solid body rotation. The instability was observed at very small tilts of the lid, corresponding to small Rossby numbers just as had been seen by McEwan (1970) and Thompson (1970).

While the experiments of McEwan (1970), Thompson (1970) and Stergiopoulos and Aldridge (1987) produced the shear instability in a cylinder through coaxial perturbation, it is also predicted that pure elliptical instability can be excited in a cylinder. As noted above, elliptical deformation of otherwise circular streamlines allows coupling between pairs of inertial waves leading to instability of the elliptic vortex. This elliptical instability was first observed by Gledzer *et al.* (1975) in a transient experiment on a fluid-filled, right cylinder of elliptical cross-section. These authors used a rigid cylinder with semi-axes 50 and 60 mm with height variable over the range 80–370 mm filled with fluid containing balls of synthetic resin which were neutrally buoyant in a slightly salted water solution. By suddenly stopping the container, the fluid was forced to rotate past a stationary elliptical boundary thus producing instability. An illustration of a three eddy instability produced is shown in Figure 4 (their Figure 1) which depicts a similarly constructed apparatus by Novikov.

Another transient experiment following the style of Gledzer *et al.* (1975) was done by Valdimirov and Tarasov (1985) with flow visualization through release of dye along the axis of a right cylinder with slightly elliptical cross-section. A photograph of the axial distortion of dye over time since the container stopped rotating is shown in Figure 5 taken from their paper. When the container stops, the fluid continues to rotate through the now stationary elliptical boundary. This elliptical strain couples inertial modes to produce elliptical instability which develops as growing wave, standing in the laboratory frame of reference.

A steady version of the above transient experiment was performed by Malkus (1989) by producing a tide-like distortion of a flexible-walled, right cylinder. This laboratory fixed, elliptical distortion of the cylinder's outer wall produced a repeated collapse of a sinuous disturbance seen forming on the axis in Figure 6 taken from Malkus (1989). Eloy *et al.* (2000) extended this experiment over a range of Ekman (they refer to Reynolds numbers) numbers and found what appear to be saturated states, as illustrated in Figure 7. These states show up as standing waves in the perturbation (laboratory fixed) frame of reference but do not collapse until the Ekman number is sufficiently small to permit instability growth. Just as in the experiments of Valdimirov and Tarasov (1985), the fluid height had to be adjusted so that resonant conditions were met as the perturbation, while still coaxial with the rotation, was necessarily equal and opposite to to the rotation speed.

Aldridge *et al.* (1997) excited the elliptical instability in a spheroidal shell by elliptically deforming a flexible inner boundary at a prescribed rotation rate. When this perturbation

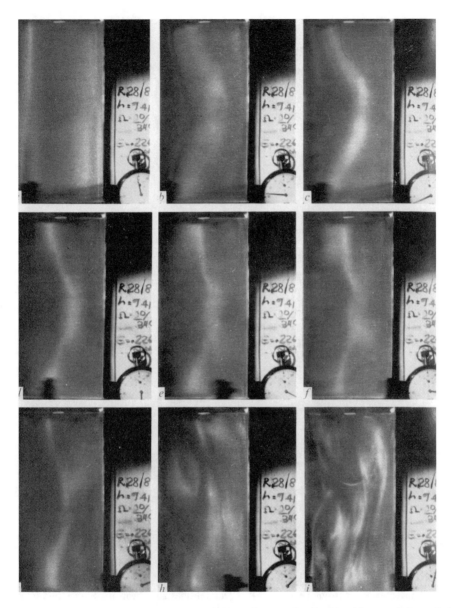

Figure 2 Time sequence demonstrating resonant collapse. (Reprinted with permission of the Cambridge University Press, from McEwan, A. D., "Inertial oscillations in a rotating fluid cylinder," *J. Fluid Mech.* **40**, 603–640 (1970).)

traveled in a retrograde sense at a speed equal to the prograde rotation, it was stationary in the laboratory frame so resembled the experimental arrangements of Malkus (1989). Since there is no dependence of eigenfrequency on aspect ratio in a spherical shell as is the case for a cylinder, it was necessary to arrange for the speed of the perturbation to be adjusted to tune in to the resonances which led to instability.

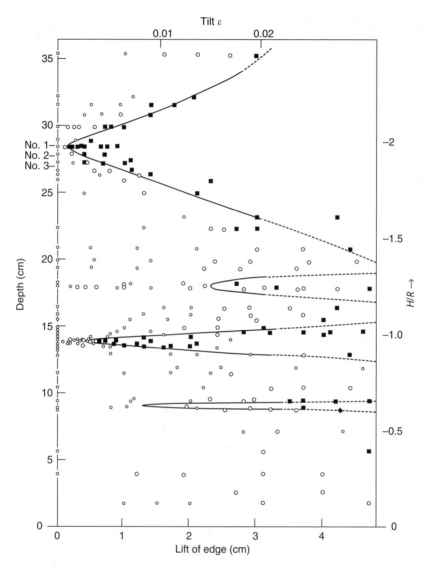

Figure 3 Shear instabilities in a rotating, tilted cylinder. (Reprinted with permission of the Cambridge University Press, from Thompson, R., "Diurnal tides and shear instabilities in a rotating cylinder," *J. Fluid Mech.* **40**, 737–751 (1970).)

A digital particle imaging velocity (DPIV) was used to make quantitative observations of the fluid's velocity field. Mounted in the rotating frame of reference was a CCD camera (512 × 480 pixels) which photographed seed particles (10 μm, neutrally buoyant spheres or aluminium flakes) over a 5 cm × 5 cm region in a plane of fixed latitude, tangent to the lower boundary of the inner sphere. Two pictures were taken at known time intervals (usually 30 ms) and particle velocities over the region were found through cross-correlation of the two images in overlapping subregions of 16 × 32 pixels. This procedure gave approximately 900 velocity vectors for the 25 cm² region.

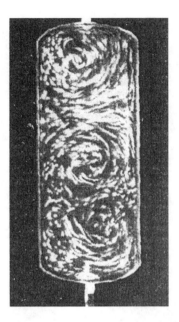

Figure 4 Three eddy instability produced by a rotating elliptical cylinder. (Reprinted with permission of American Geophysical Union, from Gledzer, E. B., Dolzhansky, F. V., Obukhov, A. M. and Ponomarev, V. M., "An experimental and theoretical study of the stability of motion of a liquid in an elliptical cylinder," *Izv. Atmos. Ocean. Phys.* **11**, 617–622 (1975), copyright by the American Geophysical Union.)

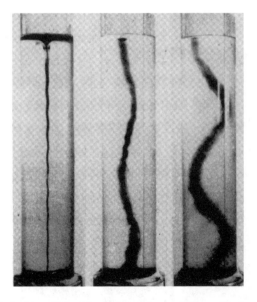

Figure 5 Transient elliptical instability. (Reprinted with permission of Springer-Verlag from figure 2 p. 721, Vladimirov, V. A. and Tarasov, V. F., "Resonance instability of the flows with closed streamlines," Ed. V. V. Kozlov, *Laminar-Turbulent Transition*, 717–722 (1985), copyright Springer Verlag.)

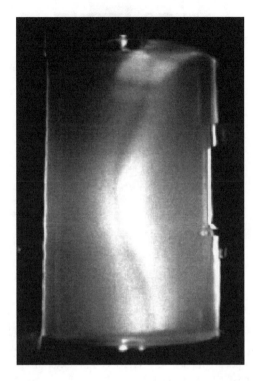

Figure 6 Elliptical instability in a cylinder, from Malkus (1989).

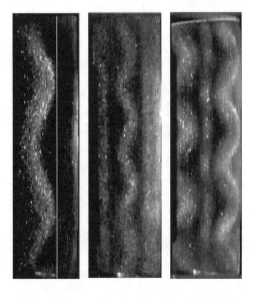

Figure 7 Saturated states from Eloy *et al*. (2000).

(a)

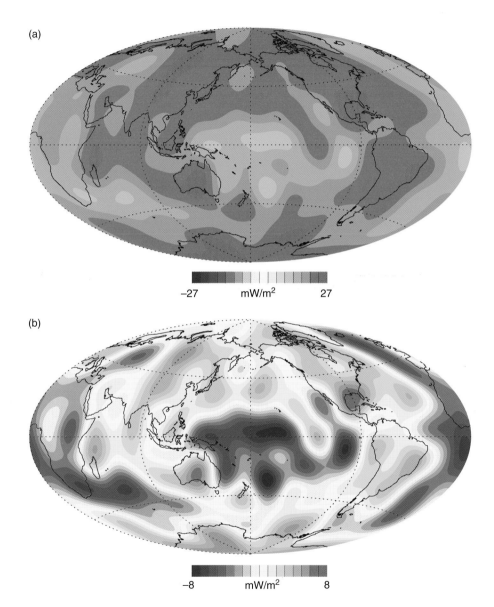

−27   mW/m²   27

(b)

−8   mW/m²   8

Colour Plate 1 Heat flow pattern on the CMB derived from a thermal boundary layer interpretation of D″ layer seismic shear wave structure shown in Figure 2, assuming 3 TW total core–mantle heat flow and a linear relationship between temperature and shear wave velocity variations. (See Chapter 1, Figure 3.)

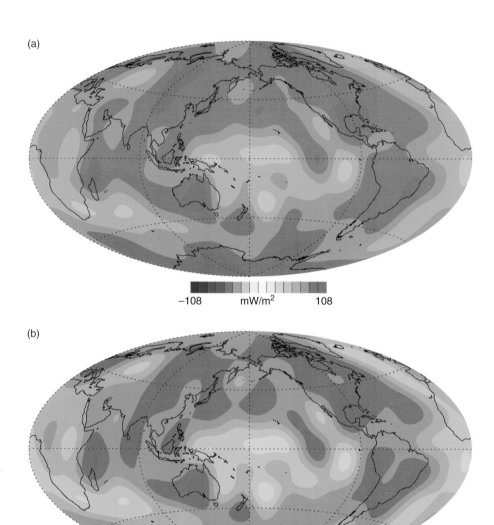

Colour Plate II Heat flow pattern on the CMB derived from a thermal boundary layer interpretation of D″ layer seismic shear wave structure shown in Figure 2, assuming 12 TW total core–mantle heat flow and a linear relationship between temperature and shear wave velocity variations. (See Chapter 1, Figure 4.)

(a)

(b)

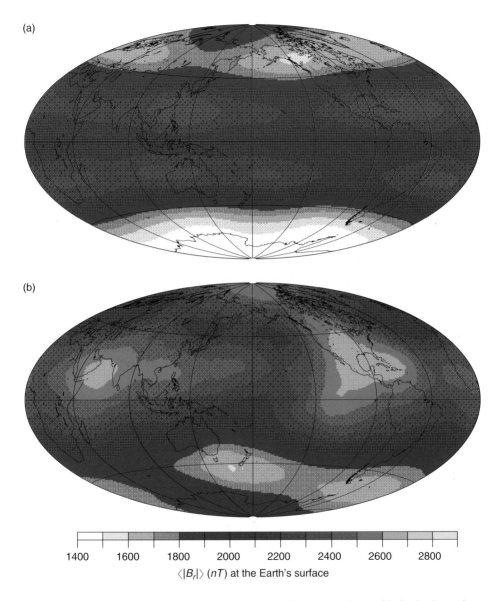

1400   1600   1800   2000   2200   2400   2600   2800

$\langle |B_r| \rangle$ (nT) at the Earth's surface

Colour Plate III Nondipole field intensity at Earth's surface averaged over 20 dipole decay times from numerical dynamo models. (See Chapter 1, Figure 8.)

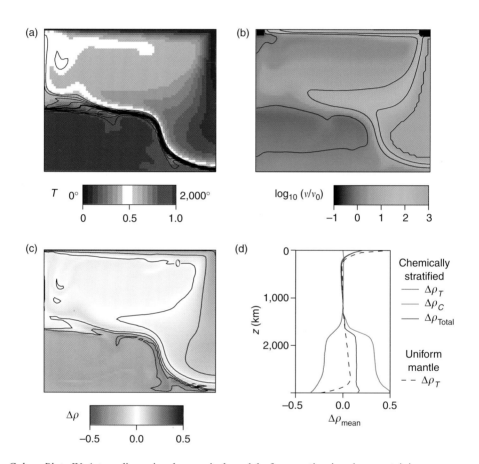

Colour Plate IV A two-dimensional numerical model of convection in a box containing a compositionally dense lower layer enriched in heat-producing elements. (See Chapter 2, Figure 2.)

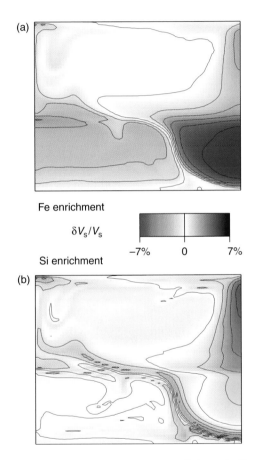

Colour Plate V Seismic velocity variations in the model of Figure 2. (See Chapter 2, Figure 3.)

Composition　　　　　　　　　Temperature (residual)

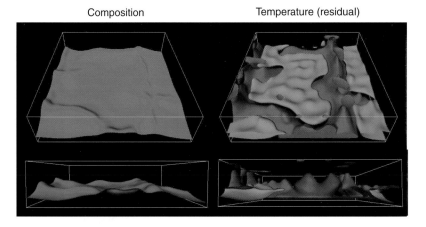

Colour Plate VI Isosurfaces of composition and residual temperature in a three-dimensional numerical model of convection. (See Chapter 2, Figure 4.)

270 km above CMB　　　770 km above CMB　　　1850 km above CMB

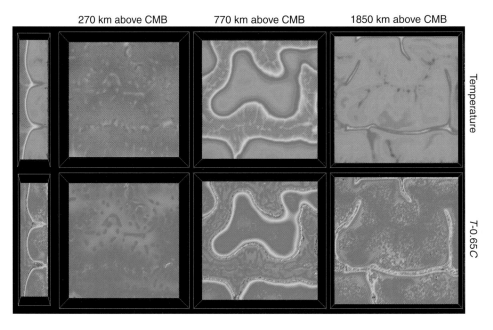

Colour Plate VII Views of the temperature field and a proxy of the seismic velocity field for the model of Figure 4. (See Chapter 2, Figure 5.)

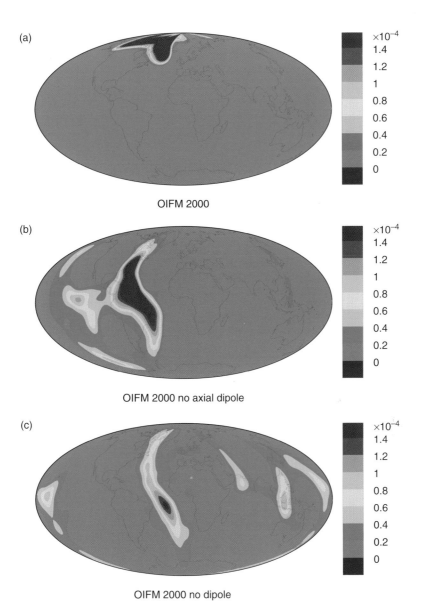

Colour Plate VIII Probability density of VGP positions for OIFM 2000 (a) includes g_1^0, (b) omitting g_1^0, and (c) omitting all dipole terms. (See Chapter 4, Figure 1.)

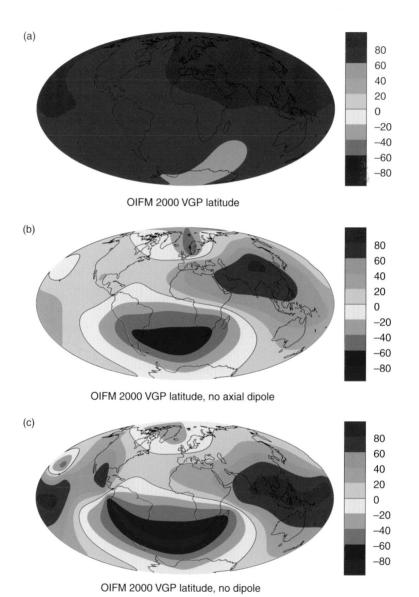

(a)

80
60
40
20
0
−20
−40
−60
−80

OIFM 2000 VGP latitude

(b)

80
60
40
20
0
−20
−40
−60
−80

OIFM 2000 VGP latitude, no axial dipole

(c)

80
60
40
20
0
−20
−40
−60
−80

OIFM 2000 VGP latitude, no dipole

Colour Plate IX Predicted VGP latitudes from OIFM 2000 (a) including g_1^0, (b) omitting g_1^0, and (c) omitting all dipole terms. (See Chapter 4, Figure 2.)

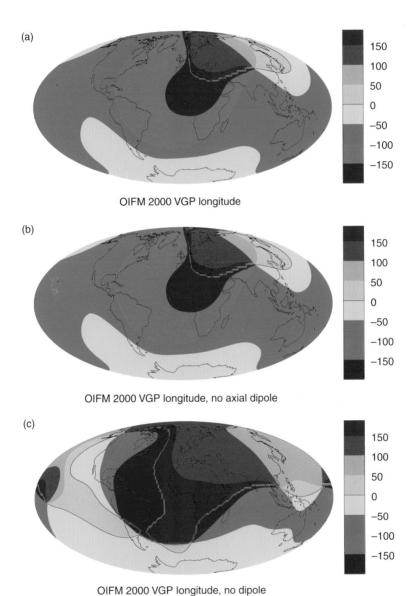

(a)

150
100
50
0
−50
−100
−150

OIFM 2000 VGP longitude

(b)

150
100
50
0
−50
−100
−150

OIFM 2000 VGP longitude, no axial dipole

(c)

150
100
50
0
−50
−100
−150

OIFM 2000 VGP longitude, no dipole

Colour Plate X Same as Figure 2, but for VGP longitude. Note that the VGP longitudes are the same whether or not the axial-dipole term is included, because it has no dependence on longitude. (See Chapter 4, Figure 3.)

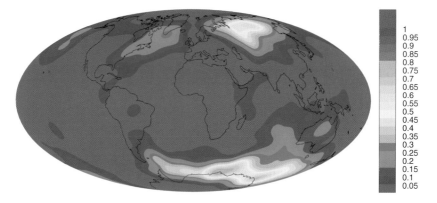

Colour Plate XI VGP density distribution for MBD97, the Matuyama–Brunhes transitional database compilation of Love and Mazaud (1997). Each direction is assigned unit weight in estimating the density function. (See Chapter 4, Figure 7.)

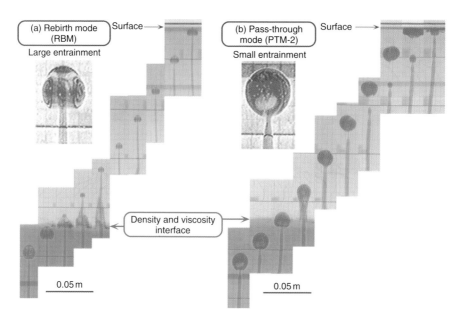

Colour Plate XII Composite images of the ascent history of a light plume across the interface between two miscible liquids. Two different behaviours are observed depending on the density and viscosity ratios: the rebirth mode (on the left) and the pass-through mode (on the right). (See Chapter 7, Figure 14.)

(a)

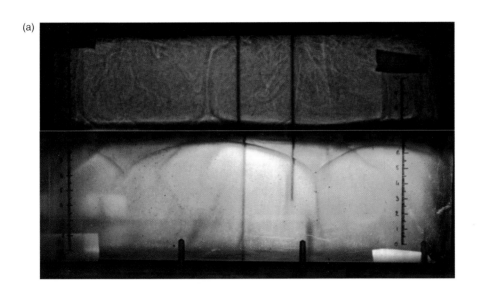

(b)

Colour Plate XIII Patterns and entrainment in thermal convection of two superposed miscible liquids. Top: the density ratio of the two liquids is large compared to that of temperature induced variations. Thin filaments are advected away from convergence wedges at the interface. Bottom: the intrinsic density ratio is similar to that of temperature induced variations. Large domes form and can reach the opposite boundary before relaxing back. (See Chapter 7, Figure 13.)

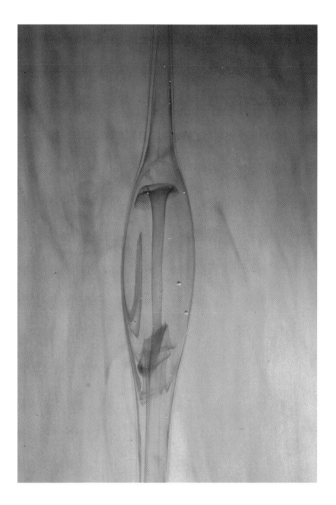

Colour Plate XIV A solitary wave propagating in the conduit of a thermal plume. A sudden increase in the influx rate of a mature plume triggers the formation of a buldge that propagates upward in a solitary wave fashion. Dye is injected in the plume for visualization. (See Chapter 7, Figure 15.)

Elliptical perturbation of the fluid will excite inertial modes of azimuthal wave number two as well as the instability of elliptical origin. Several inertial modes were observed and compared with predicted velocity fields obtained from a finite element calculation by Henderson (1996). Since there are no analytical solutions for inertial modes in a spherical shell, only approximate representations are possible. Depicted in Figure 8 are the observed (left) and predicted (right) velocity fields for the (4,1,2) mode in a spherical shell from Seyed-Mahmoud (1999). Two sets of comparisons are shown here because the mode is travelling azimuthally and has been captured at two different phases. Similarity of the velocity fields and the eigenfrequencies between observation and prediction confirms the existence of the mode. Differences between the observation and prediction are due to the presence of neighbouring modes in the observations as well as the fact the prediction is for the inviscid case.

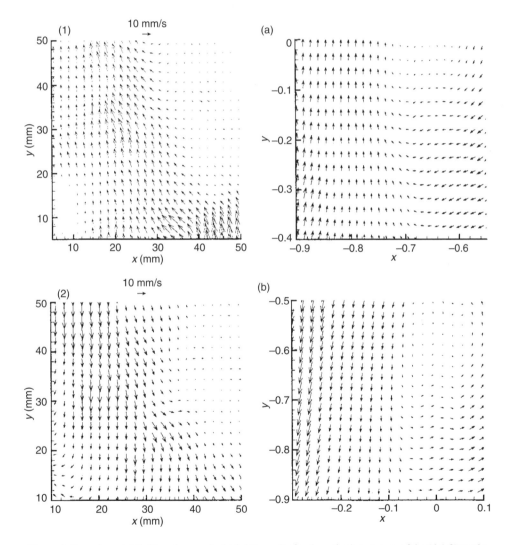

Figure 8 Experimental (left) and numerical (right) results for the velocity vectors of the (4,1,2) mode; $\lambda_{exp} \approx 1.260$, $\lambda_{num} \approx 1.247$; $\Omega = 3.239$ rad/s, $\omega = 4.082$ rad/s.

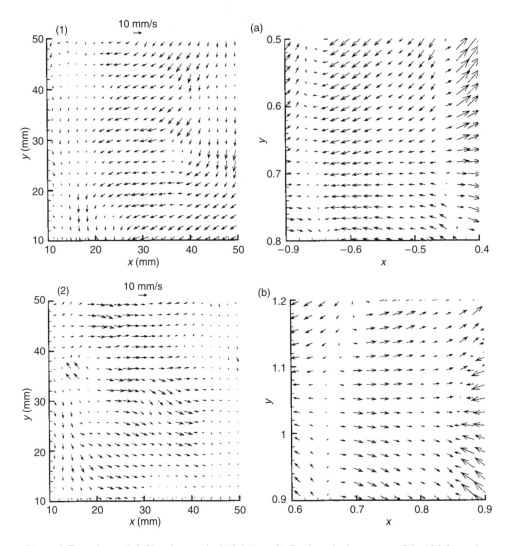

Figure 9 Experimental (left) and numerical (right) results for the velocity vectors of the (6,2,2) mode; $\lambda_{exp} \approx 0.897$, $\lambda_{num} \approx 0.896$; $\Omega = 3.325$ rad/s, $\omega = 2.983$ rad/s.

A second mode, the (6,2,2) is shown Figure 9 where the structure of the velocity field is more complex.

Velocity fields for the spin-over instability have been calculated by Seyed-Mahmoud *et al.* (2000) using a Galerkin method. The functions used to represent the instability were approximate inertial modes obtained as described above. Depicted in Figure 10 is a comparison of observation (left) and prediction (right) for the spinover instability based on the (2,1,1) mode. Three pairs of observations and prediction are shown at different phases of the growth since the instability is almost stationary in the laboratory frame of reference, but of course travelling azimuthally in the camera's (rotation) frame of reference.

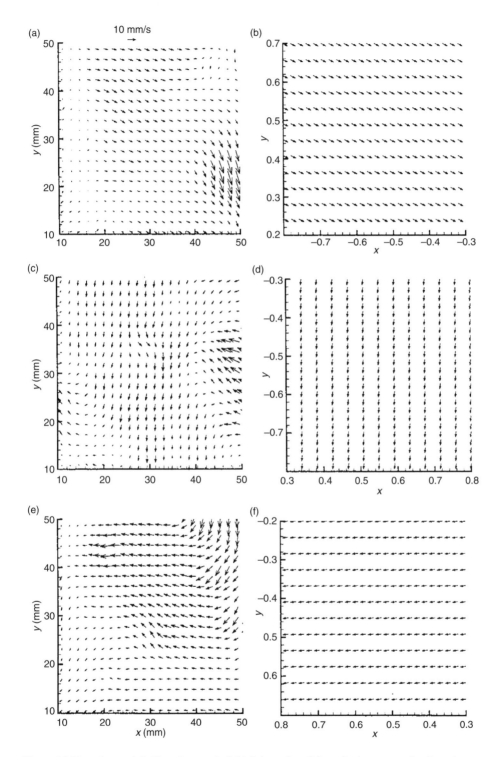

Figure 10 Experimental (left) and numerical (right) results of the velocity vectors for the spin-over instability captured; $\omega/\Omega \approx 0.950$; $\Omega = 3.305\,\text{rad/s}$.

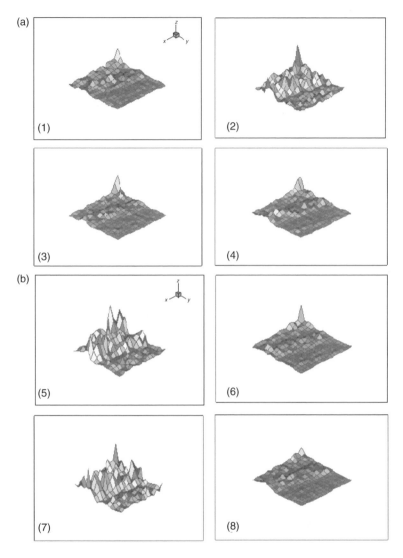

Figure 11 Onset and decay of instability in a spherical shell as seen in wavenumber space; interval between images approximately 90 s; $\omega/\Omega = 0.93$, $\varepsilon = 0.09$.

While the perturbation continues the instability decays and then subsequently grows again. This sequence continues and has been captured in Figure 11, which displays snapshots of the two-dimensional spatial Fourier transform of the region of observation. Growth and decay of the instability from one frame to the next is clearly visible, much like as had been observed by Malkus (1989) with his continuing external, tide-like perturbation of the flexible cylinder.

3.2. Oblique axes

In this section, we group together the observations of elliptical/shear instability produced in a contained rotating fluid by perturbing the fluid about an axis other than the axis of rotation.

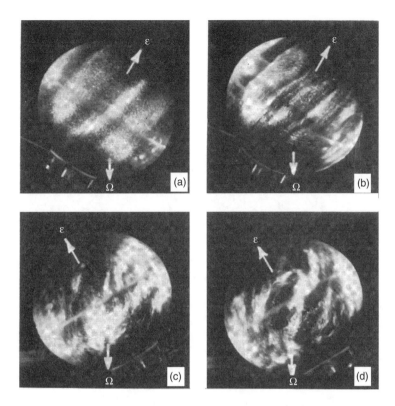

Figure 12 Instability in a precessing ellipsoid. (Reprinted with permission from Malkus, W. V. R., "Precession of the Earth as the cause of geomagnetism," *Science* **160**, 259–264 (1968). Copyright 1968, American Association for the Advancement of Science.)

Precession of a rotating container accomplishes this, as an interpretation by Kerswell (1993) of the experiments of Malkus (1968) has demonstrated. Shown in Figure 12 from Malkus (1968) are cross-sections of the precessing spheroid of ellipticity $\eta = (a - c)/c = 1/24$, rotating at rate $\omega_c = 60$ rpm and containing fluid seeded with aluminium flakes illuminated by a sheet of light.

Three distinct regimes of flow were observed, depending on the retrograde precession rate, Ω_p. Panel (a) corresponding to $\Omega_p = -0.75$ rpm shows laminar flow while panel (b) at $\Omega_p = -1.0$ rpm already shows the wave like disturbances seen in the cylindrical geometry above. Panels (c) and (d) at $\Omega_p = -1.33$ rpm show the turbulent regime. Although both shear and elliptical distortion of streamlines is produced by precession of a spheroid, Kerswell (1993) calculates that shear predominates and hence leads to instability in the Malkus experiment. Figure 13, derived from strains calculated for a precessing spheroid by Kerswell (1993), illustrates this predominance as well as the asymmetry between prograde and retrograde precession with respect to strain for the flattening used by Malkus.

For the Poincaré number P = precession rate/rotation rate, the two cases, $P = 0.016$ and $P = 0.022$ correspond respectively to the cases (b) and (c) in Figure 12, cited above. Note that the shear strain exceeds the elliptical strain and it will be seen below that at very low values of Poincaré number corresponding the Earth's fluid outer core, shear greatly exceeds elliptical

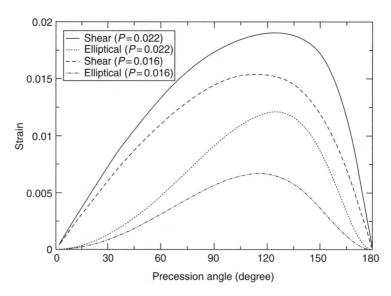

Figure 13 Strain rates calculated from Kerswell (1993) for a spheroid of flattening $\eta_k = a^2 - c^2/a^2 = 49/576$ used by Malkus (1968). The Poincaré numbers shown are for the two cases on opposite sides of the transition to turbulence. Prograde ($\leq 90°$) and retrograde ($\geq 90°$) precession are shown; Malkus (1968) experiment was done at 150°.

strain. The readily apparent asymmetry with respect to precession angle 90° at the relatively large value of flattening of the Malkus (1968) experiment means that the strain is much higher for a retrograde precession than for a prograde precession of equivalent magnitude. This may explain the observations in experiments on precession that flows for retrograde precession are more readily unstable than for prograde precession.

What appears to have been inertial instability of shear type was observed by Gans (1970) in a rapidly rotating cylinder which was precessed at an angle of 90°. He first confirmed prediction from linear theory by observing steady flow corresponding to standing inertial waves in the precessing frame of reference by using a highly viscous glycerine–water solution as the working fluid. No instability was seen, presumably due to the large dissipation. When this fluid was replaced with water, however, he found what he described as 'pitching about' on the axis and a then a collapse as had been seen by McEwan (1970) and reported more recently by Malkus (1989). Apparently the shear instability was only able to grow when the Ekman number was sufficiently small, as was the case with water as the working fluid.

Some closely related pitching to that observed by Gans (1970) was found during the late 1970s when a practical problem arose concerning the flight instability of a spin-stabilized, liquid-filled, projectile. The shell, filled with white phosphorus, was designed as a battlefield obscurant by the US military. The phosphorus melted when ambient conditions became sufficiently hot, and modes of oscillation were excited in the shell at a frequency near the shell's nutational frequency which led to instability of the shell during flight. One example from a series of laboratory experiments designed to study the stability of a liquid filled, precessing cylinder is shown from work by Miller (1981) in Figure 14.

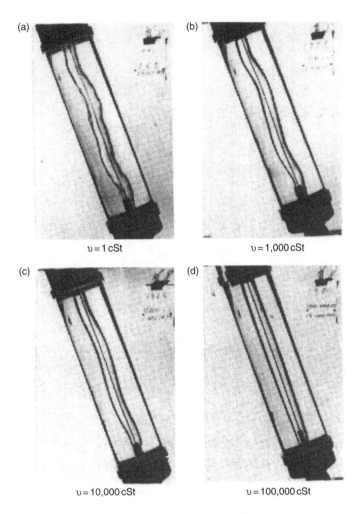

Figure 14 Effect of dissipation on saturated state in a precessing cylinder, from Miller (1981). Viscosities are 1 cSt (top, left), 1,000 cSt (top, right), 10,000 cSt (bottom, left) and 100,000 cSt (bottom, right).

Depicted in this figure is a partially filled, right circular cylinder, spinning at 4000 rpm and precessed in a prograde sense at 400 rpm with an angle of 20° between the two axes. Viscosity increases from 1 to 100,000 cSt from top left to bottom right. The central line is a reference wire stretched along the cylinder's axis. The observed sinusoidal distortion appears to be a saturated state as seen by Eloy *et al.* (2000) in their observations of an elliptical strain of streamlines in a rotating cylinder. In the Miller (1981) observation, there is a progressively larger amplitude of the wave, standing in the precession frame, as the Ekman number decreases, to the point of irregularity seen in the waveform at the smallest viscosity.

Also using a precessing cylinder but at very small amplitudes, Manesseh (1992, 1994) appears to have found this instability as illustrated in Figure 15. The sequence of panels in time shows the development of the instability, its decay and finally the re-establishment of the laminar regime while the precession continues. This is a similar response to the one seen

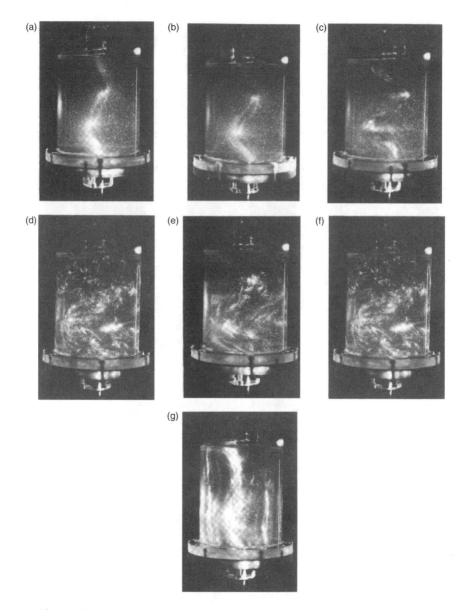

Figure 15 Instability in a retrograde precessing cylinder; the tilt of the cylinder is 177 (3)°; Poincaré number, $P = 0.355$. (Reprinted with permission of the Cambridge University Press, from Manasseh, R., "Breakdown of inertia regimes in a precessing cylinder," *J. Fluid Mech.* **243**, 261–296 (1992).)

later by Malkus (1989) and Aldridge *et al.* (1997) where the perturbation continues while the instability grows, decays and forms again. In both cases the growing disturbance is stationary in the perturbation frame. In a set of experiments with increasing Ekman number, Manasseh (1992) also noted a small increase in the time taken for the flow to become unstable. This is

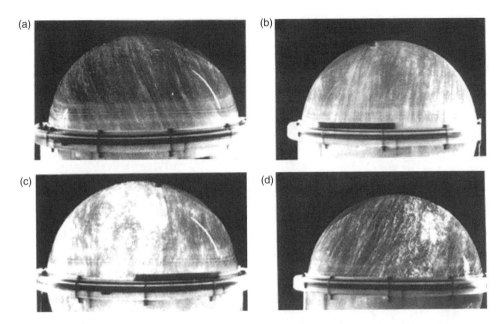

Figure 16 Instability in a precessing spheroid; precession angle $\alpha = 156.5°$ (23.5°), $P = 0.0058$ (top), $P = 0.058$ (bottom). (Reprinted with permission of Blackwell Publishing from Vanyo J., Wilde P., Cardin P. and Olson, P. "Experiments on precessing flows in the Earth's liquid core," *G. Journal Int.* **121**, 136–142 (1995), copyright Blackwell Science Ltd. & Geophysical Journal International.)

the behaviour expected for the growth rate of inertial instability, which should decrease as Ekman number increases, as discussed in Section 4.3.

Precession of a water-filled oblate spheroid with flattening 1/100, was reported by Vanyo *et al.* (1995) and more recently with flattening 1/400 by Vanyo and Dunn (2000). In the earlier experiments, from which some results are shown in Figure 16, a transition from laminar to turbulent flow was found at precession speeds and amplitudes consistent with onset of inertial instability. The reflecting aluminum flakes clearly show turbulent flow at the larger value of Poincaré number, P. Specifically Vanyo *et al.* (1995) found that the ratio of growth rate to dissipation obtained from Kerswell (1993) was close to unity for their experimental conditions corresponding to the onset of turbulence. It is worth noting here that only shear strain was used in this calculation and if the elliptical strain which is almost half of the shear strain, is included and the unstable response was stationary for both types of strain, even better agreement between experiment and the instability model would have been found. It is also noteworthy that Vanyo *et al.* (1995) concluded that application of this albeit highly idealized instability model using flattening, Poincaré and likely Ekman numbers of the fluid core could not be ruled out as insignificant with respect to its role in core dynamics.

More recently, Noir (2000) precessed a rotating spheroid of fluid using the same ellipticity $\eta = 1/24$ as that of Malkus (1968). Noir observed the transition to turbulence as a function of the Poincaré number, P for a precession angle of $\alpha = 20°$. Noir (2000) found as Vanyo *et al.* (1995) and Vanyo and Dunn (2000) did, that prograde precession was more stable than retrograde precession. These observations are illustrated in Figures 17 and 18 for the

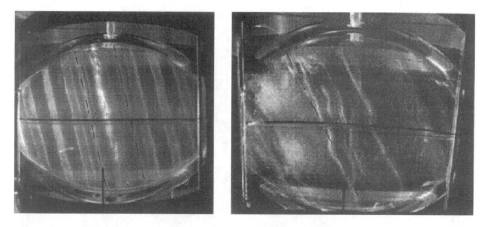

Figure 17 Prograde precession of a spheroid, from Noir (2000). Poincaré number $P = 0.033$, precession angle, $\alpha = 20°$ (left); $P = 0.1, \alpha = 33°$ (right).

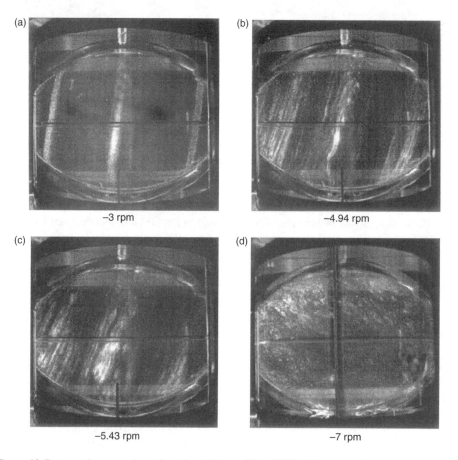

(a)   −3 rpm

(b)   −4.94 rpm

(c)   −5.43 rpm

(d)   −7 rpm

Figure 18 Retrograde precession of a spheroid, from Noir (2000). For precession angle $\alpha = 160°$ (20°), clockwise from top, left, $P = 0.01, 0.016, 0.018$ and 0.023.

prograde and retrograde cases, respectively. In the prograde case, the flow remains almost laminar at $P = 0.033$ while in the retrograde case the fluid is fully turbulent at $P = 0.023$ for the same magnitude of precession angle. But note that there is evidence of the familiar 'S' shape on the axis for the onset of the inertial instability in both figures, suggesting that the observed instability originates in the fluid interior rather than in the boundary layer. As seen above in Figure 13, the large asymmetry in strain between the two cases $\alpha = 20°$ (prograde) and $\alpha = 160°$ (retrograde) might account for the observed asymmetry of stability between these two cases.

4. Applications to the Earth's core

Observations of the Earth's gravity and magnetic fields are related to current models for their assessment. Successful models can then provide estimates of Earth parameters that determine the models. Here, three sets of observations are presented and linked to new models of core dynamics. First, the polar vortex found by Olson and Aurnou (1999) is linked to geostrophic flow associated with mantle precession as described by Noir *et al.* (2001). Second, very long records from several superconducting gravimeters, combined and inverted by Smylie *et al.* (2001), give new estimates for the viscosity of the fluid outer core near the inner core, outer core boundary as well as periods for the Slichter modes. Third, paleomagnetic intensity data derived from sedimentary cores are inverted to give time scales that can be associated with growth and decay of the geomagnetic field.

4.1. Polar vortex and geostrophic flow

Olson and Aurnou (1999) have found what they describe as a polar vortex to account for the westward drift of the geomagnetic field of about 180° in the interval 1870–1990, illustrated in the left hand panels of Figure 19 from their paper. Although no specific process is cited, reference is made vortices generated in rotating fluids. One such vortex is that associated with a global geostrophic flow due precession of the Earth's mantle. It has been estimated by Noir *et al.* (2001) that the magnitude of such a flow in the boundary layer near the critical latitude, is $\gamma^2 E^{-3/10}$ based on his precession experiments and a non-linear model. Here $\gamma = 2.62 P/\eta$ is a parameter based on the Poincaré number where $\eta = 1/400$ is the flattening of the CMB. If the polar vortex is indeed a result of geostrophic flow resulting from precession of the mantle, equating the observed rate given above to the above geostrophic magnitude, leads to a value of $E \approx 10^{-11}$ for the core. Only order of magnitude estimates of geostrophic flows have been made here since the neglect of both buoyancy and Lorentz forces in the instability model do not justify a more precise calculation. Perhaps more importantly, however, the peak value of retrograde drift of the polar vortex found by Olson and Aurnou (1999) was at a colatitude of about 10° while that found in the precession experiments of Malkus (1968) and in the calculations of Noir *et al.* (2001) gave a peak retrograde azimuthal drift at 7.5° colatitude.

4.2. Superconducting gravimetry and core viscosity

Measurements of gravity using stable SGs over several years at the Earth's surface can reveal temporal changes in gravity due to any fluctuations in the density distribution within the Earth. Smylie *et al.* (2000) have combined such long data sets from Vienna from (2/8/95 to 17/11/99; 37,623 h), Membach from (4/8/95 to 7/10/99; 36,620 h) and Boulder from (12/3/95 to 12/11/99; 30,000 h) by a product spectrum method. The resulting spectra as well as those obtained from earlier data sets (Courtier *et al.*, 2000) are plotted in Figure 20.

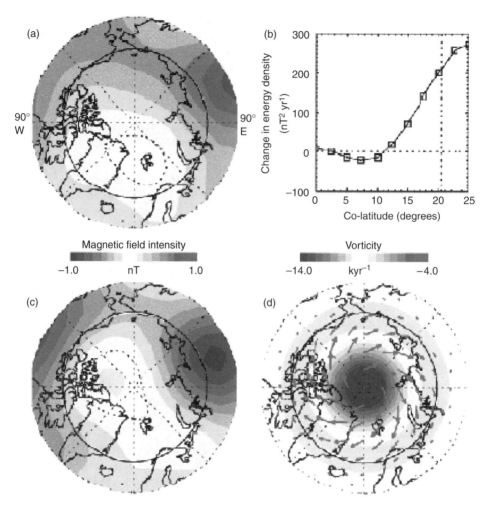

Figure 19 Polar vortex inferred from drift of geomagnetic field. (Reprinted with permission of Nature from Olson, P. and Aurnou, J. "A polar vortex in the Earth's core," *Nature* **402**, 170–173 (1999).)

The Slichter triplet of resonances are identified by their strict adherence (to parts in 10^4) to a splitting law obtained from elementary dynamics. A computer-based search of the complete spectrum found no other correctly split triplets of resonances, and a statistical analysis showed that there is only one chance in 6.8×10^{38} of obtaining a better fitting triplet on a purely random basis (Smylie *et al.*, 1993). It is important to note, however, that the product spectral method does not use phase information since it combines only amplitudes across stations. A global phase experiment conducted by Courtier *et al.* (2000) yields central frequencies which differ from those found previously by 0.092, 0.064 and −0.062% for the retrograde, axial and prograde modes, respectively.

Recent results of Smylie *et al.* (2001) are shown in Table 2. The first row gives the observed periods using the product spectrum method for the four stations as described above. Using

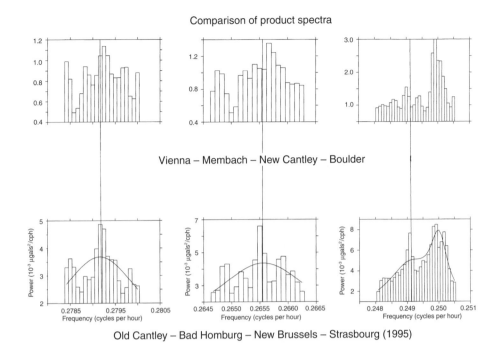

Figure 20 Product spectra of superconducting gravimetry data, from Smylie *et al.* (2000) and Courtier *et al.* (2000); from recent data sets (upper) (Smylie *et al.*) and earlier data sets (lower) (Courtier *et al.*). Lines show Slichter frequencies identified by their strict adherence to the splitting law. The Solar atmospheric tide at 0.25 cpd is clearly evident in the right hand panels.

Table 2 Observed and fitted Slichter periods and viscosities, from Smylie *et al.* (2001)

	Retrograde (h)	*Axial* (h)	*Prograde* (h)
Observed periods	3.5822 ±0.0012	3.7656 ±0.0015	4.0150 ±0.0010
CORE11 viscous periods ($\nu = 9.58 \times 10^6$ m^2/s)	3.5793	3.7647	4.0121
Cal8 viscous periods ($\nu = 10.21 \times 10^6$ m^2/s)	3.5840	3.7731	4.0168
Cal8 inviscid periods	3.5168	3.7926	4.1118

Ekman boundary layer theory, they have calculated the Slichter periods and viscosities near the ICB for CORE11 and Cal8 core models which are given in rows two and three while the inviscid results for Cal8 are given in row 4 for comparison.

The new data sets with wide geographical distribution employed by Smylie *et al.* (2001) show the previously recovered translational mode signals in the product spectrum. In addition, reduction in the rotational splitting of the equatorial translational modes provides a viscosity measurement method with built-in redundancy and spectroscopic precision. The recovered viscosity of $v = 10.21 \times 10^6\,\mathrm{m}^2/\mathrm{s}$ appears to confirm the semi-fluid nature of the F-layer suggested by Braginsky as the source of energy for the geodynamo. While no parallel theory exists as yet for shear viscosity, the value is near that at low frequencies for bulk viscosity obtained by Stevenson (1983). The Cal8 earth model gives an axial mode period within 0.2% of the observed value. Since this mode is little affected by rotation or viscosity this Earth model is in remarkable agreement with observation. The modal periods are extremely sensitive to inner core density giving a resolution of a few parts in 10^5.

4.3. Inversion of paleomagnetic intensity data

If an inertial instability has existed in the Earth's core, it would not be surprising to find evidence of it in the temporal signature of the geomagnetic field. Even if the geodynamo is driven by a convective source of energy, the onset of a rotational instability as seen in the laboratory experiments described above, could produce a turbulent disruption of the velocity field driving the dynamo. Accordingly, temporal behaviour of the geomagnetic field on time scales associated with the instability may exist in paleomagnetic intensity data.

Records of paleomagnetic intensity for sedimentary deposits have become available and an example is shown in Figure 21 from Channell [personal communication (1999)]. Immediately obvious from this figure is the significant temporal behaviour in amplitude of the field while it remains in its 'normal' state up to the time of the MB reversal just before 800 ka ago. Although there are significant uncertainties in the time-base of this record due to variations in sedimentation rates as well as random errors in the dating process, some measure of the time dependence of the paleomagnetic intensity is available. A zoom of the time sequence

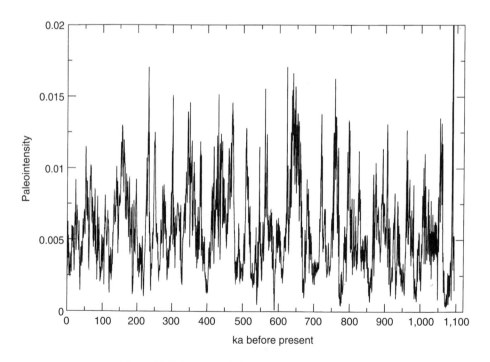

Figure 21 Paleomagnetic intensity data from OPD983.

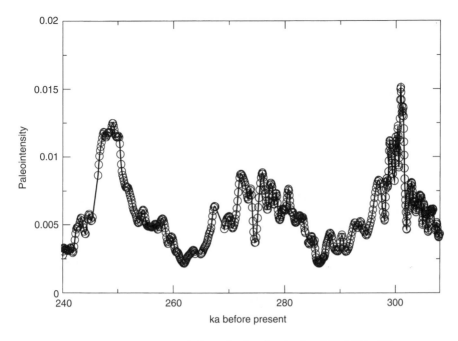

Figure 22 Paleomagnetic intensity data for the time 240–308 ka BP.

of intensities is shown in Figure 22 where the circles represent the paleointensities and the lines shown simply connect the data points.

The data in this figure can be considered as three growths and decays of the intensity where real time runs from right to left. The model here is that the field is going through growth and decay as has been observed in the growth and decay of the instability in the laboratory experiments. Clearly this is a drastic oversimplification of actual core processes, but only the time dependence is needed here.

Each of the six segments was modelled as an exponential and the resulting fits are shown in Figure 23. Above each panel is the e-folding time for the fitted decay or growth.

The fitting procedure was repeated over the entire record of 1,100 ka. Decay and growth rates found are plotted in Figure 24, which shows all rates which correspond to e-folding times longer than 4 ka, since those few cases (5/74) of very short decays or growths are unphysical due to screening of the mantle. Although there are significant fluctuations in the plotted rates, there appears to be a trend towards larger decays and growths at more recent times.

In order to relate the time scales observed to those of possible inertial instabilities in the core, the conditions for the existence of these instabilities are reviewed. Two inertial modes can be coupled through either elliptical deformation of streamlines if their azimuthal wavenumbers differ by 2 or by shearing strain of streamlines if their azimuthal wavenumbers differ by 1. Instability can occur if the growth rate is greater than its dissipation rate, namely

$$\alpha\varepsilon - \beta\sqrt{E},$$

where ε = deformation or strain and E = Ekman number, and both α and β are typically of O(1) (Kerswell, 1993).

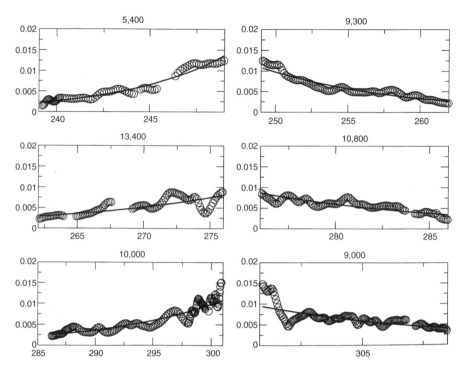

Figure 23 Exponential fits to data of Figure 22. Numbers above each graph are the e-folding times in years.

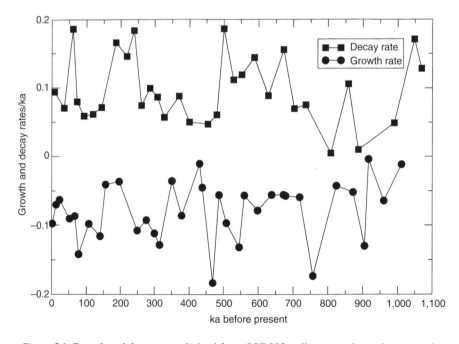

Figure 24 Growth and decay rates derived from OPD983 sedimentary data using regression.

Estimates for *inviscid* instability growth rates in the core from elliptical deformation, $\varepsilon \approx 8.5 \times 10^{-8}$, of the CMB due to the semi-diurnal tide is

$$0.5\varepsilon\Omega \approx 3.1 \times 10^{-12}\,\mathrm{s}^{-1} \quad \text{or} \quad T_\mathrm{e} \approx 10{,}000 \text{ years.}$$

Shear strain, $\varepsilon \approx 4 \times 10^{-8}$, due to the 26,000-year precession is

$$0.6\varepsilon\Omega \approx 1.7 \times 10^{-12}\,\mathrm{s}^{-1} \quad \text{or} \quad T_\mathrm{s} \approx 18{,}000 \text{ years.}$$

These e-folding times are of course minimum times since they will be extended when growth rates are reduced due to dissipation in the core. While these times are comparable to the reciprocals of the decays shown in Figure 24, uncertainties in the time base of the paleomagnetic data as well as fluctuations in the rates themselves preclude quantitative determinations of the Ekman number E for the core. Also complicating a determination of E is the fact that our drastically simplified model ignores Lorentz (Laplace) forces. Indeed, the ohmic decay time of the field is of the same order as the values recovered from the paleomagnetic data so that rates recovered may be directly determined by both processes. Since the growth rates recovered are of the same order as decay rates, it seems likely that a process more complex than diffusive decay has been found.

4.3.1. Polarity superchrons

The existence of inertial parametric instability in a fluid depends on the inviscid growth rate exceeding the dissipation rate. In application of this point to the Earth's core, it is necessary to consider very long time-scales since the growth rates are determined by the very small tidal and precessional amplitudes. Indeed, because of this fact, it is possible to extract rotational energy on a diurnal time scale and accumulate it over a geomagnetic time-scale. Accordingly, changes in core properties on geomagnetic time-scales will determine what becomes of the instability. While rotation rate of the Earth has changed over geological time due to tidal friction as well as the tidal amplitudes themselves, it seems likely that thermal changes over geological time could significantly alter the dissipation rate. Thus extended periods of geological time, corresponding to the polarity superchrons, could pass with the dissipation rate of instabilities exceeding their growth rate.

5. Conclusions

Renewed interest in parametric instabilities of a rotating fluid has led to the re-interpretation of several experiments as described in Section 2. The ubiquitous presence of elliptical and shear instabilities in experiments on contained rotating fluids requires a reconsideration of the assumptions made by Loper (1975) and Rochester *et al.* (1975) who ruled out the Earth's precession as a driving mechanism for the geodynamo. Since those authors did not consider parametric instability in their otherwise careful analysis of the possible role of precession in dynamo action, tidal forces now need to be reconsidered once more. Numerical models of the geodynamo by Glatzmeier and Roberts (1995), for example, who have successfully produced working dynamos which are convection driven, have not included rotationally instabilities. For the relatively large values of Ekman number in their models, there would of course be no inertial instability since dissipation rates would swamp their growth. If, however, more likely values of Ekman number for the core were used, boundary layer regions between the interior Poincaré flow and the CMB might prove interesting in the maintaining the geodynamo. Indeed one interpretation of the paleomagnetic intensity data described in Section 4.3

suggests that growth and decay of the geomagnetic field could be produced by a rotational instability.

Renewed interest in tidal forces and precession has been significant in recent time, both through further geophysical interpretation of earlier experiments of Vanyo *et al.* (1995) by Pais and Le Mouël (2001) and new experiments by Noir (2001). We are beginning to understand the complex problem of the transition from laminar to turbulent flow in precession through quantitative observations in experiments and their interpretation incorporating earlier theoretical models of Busse (1968) and Kerswell (1993). Instability arising from both boundary layer shear as well as internal shear and elliptical instability appear to important in the transition to turbulent flow. But it is the understanding of this phenomenon through the interpretation of laboratory experiments which will ultimately allow us to decide what role an inertial instability or boundary layer shear might have in core dynamics.

Observations of changes in the Earth's gravity field over time using SGs from some dozen stations around the globe are currently being collected. At present the very small signals that have been attributed to the Slichter modes by Smylie *et al.* (2000) need further confirmation. The Chandler and annual wobble of the mantle has been detected using SG by change in latitude of an observatory simultaneously with the expected change in gravity derived from the observation of the observatory's latitude change using VLBI as demonstrated by Richter (1990). Short-period wobble, possibly associated with core oscillations through conservation of equatorial components of angular momentum could be found using many SG observatories in the same manner as demonstrated by Aldridge and Cannon (1993) using one observatory.

Acknowledgements

Support for this work came from a research grant awarded by the Natural Science and Engineering Research Council of Canada.

Permission to reproduce figures in this chapter is acknowledged from the following publishers: American Geophysical Union (Figure 4), American Institute of Aeronautics and Astronautics, Inc. (Figure 14), Blackwell Science Ltd., (Figure 16), Cambridge University Press (Figures 1, 2, 3, and 15), Nature (Figure 19), Science (Figure 12), Springer-Verlag (Figure 5).

References

Aldridge, K. D. and Cannon, W. H., "A search for evidence of short period polar motion in VLBI and supergravimetry observations," In: *Dynamics of the Earth's Deep Interior and Earth Rotation*, (Eds J.-L Le Mouël, D.E. Smylie and T. A. Herring, American Geophysical Union, pp. 17–24 (1993).

Aldridge, K. D. and Toomre, A. "Axisymmetric inertial oscillations of a fluid in a rotating spherical container," *J. Fluid Mech.* **37**, 307–323 (1969).

Aldridge, K. D., Seyed-Mahmoud, B., Henderson, G. A. and van Wijngaarden, W., "Elliptical instability of the Earth's fluid core," *Phys. Earth Planet. Inter.* **103**, 365–374 (1997).

Busse, F. H., "Steady fluid flow in a precessing spheroidal shell," *J. Fluid Mech.* **33**, 739–751 (1968).

Courtier, N., Ducarme, B., Goodkind, J., Hinderer, J., Imanishi, Y., Seama, N., Sun, H., Merriam, J., Bengert, B. and Smylie, D. E., "Global superconducting gravimeter observations and the search for the translational modes of the inner core," *Phys. Earth Planet. Inter.* **117**, 3–20 (2000).

Creager, K. C., "Inner core rotation from small-scale heterogeneity and time-varying travel times," *Science* **278**, 1284–1288 (1997).

Eloy. C., Le Gal, P. and Le Dizès, S., "Experimental study of the multipolar vortex instability," *Phy. Rev. Lett.* **85**, 3400–3403 (2000).

Gans, R. F., "On the precession of a resonant cylinder," *J. Fluid Mech.* **41**, 865–872 (1970).

Glatzmaier, G. A. and Roberts, P. H., "A three-dimensional self-consistent computer simulation of a geomagnetic field reversal," *Nature* **377**, 203–209 (1995).

Gledzer, E. B., Dolzhansky, F. V., Obukhov, A. M. and Ponomarev, V. M., "An experimental and theoretical study of the stability of motion of a liquid in an elliptical cylinder," *Izv. Atmos. Ocean. Phys.* **11**, 617–622 (1975).

Greenspan, H. P., *The Theory of Rotating Fluids*, Cambridge University Press, Cambridge (1968).

Gubbins, D., "Rotation of the inner core," *J. Geophys. Res.* **86**, 11695–11699 (1981).

Henderson, G. A., "A finite-element method for weak solutions of the Poincaré problem," PhD Thesis, York University, Toronto, Canada (1996).

Kerswell, R. R., "The instability of precessing flow," *Geophys. Astrophys. Fluid Dynam.* **72**, 107–114 (1993).

Kerswell, R. R. and Malkus, W. V. R., "Tidal instability as the source for Io's magnetic signature," *Geophys. Res. Lett.* **25**, 603–606 (1998).

Loper, D., "Torque balance and energy budget for the precessionally driven dynamo," *Phys. Earth Planet. Inter.* **11**, 43–60 (1975).

Lumb, L. I., Aldridge, K. D. and Henderson, G. A., "A generalized 'core resonance' phenomenon: Inferences from a Poincaré core model," In: *Dynamics of the Earth's Deep Interior and Earth Rotation* (Eds J. L. Le Moüel, D. E. Smylie and T. A. Herring) Am. Geophys. Union, Geophys. Monogr. **72**, 51–68 (1993).

McEwan, A. D., "Inertial oscillations in a rotating fluid cylinder," *J. Fluid Mech.* **40**, 603–640 (1970).

Malkus, W. V. R., "Precession of the Earth as the cause of geomagnetism," *Science* **160**, 259–264 (1968).

Malkus, W. V. R., "An experimental study of global instability due to tidal (elliptical) distortion of a rotating elastic cylinder," *Geophys. Astrophys. Fluid Dynam.* **48**, 123–134 (1989).

Malkus, W. V. R., "Energy sources for planetary dynamos," In: *Lectures on Solar and Planetary Dynamos* (Eds M. E. Proctor and A. D. Gilbert), Cambridge University Press, pp. 161–179 (1994).

Manasseh, R., "Breakdown of inertia regimes in a precessing cylinder," *J. Fluid Mech.* **243**, 261–296 (1992).

Manasseh, R., "Distortions of an inertia waves in a rotating fluid cylinder forced near its fundamental mode resonance," *J. Fluid Mech.* **265**, 345–370 (1994).

Miller, M. C., "Void characteristics of a liquid-filled cylinder undergoing spinning and coning motion," *J. Spacecr. Rockets* **18**, 286–288 (1981).

Noir, J., "Écoulements d'un fluide dans une cavité en précession," PhD thèse, Université Grenoble 1, France, November (2000).

Noir, J., Jault, D. and Cardin, Ph., "Numerical study of the motions within a precessing sphere, at low Ekman number," *J. Fluid Mech.* **437**, 283–299, (2001).

Olson, P. and Aurnou, J., "A polar vortex in the Earth's core," *Nature* **402**, 170–173 (1999).

Pais, M. A. and Le Mouël, J. L., "Precession-induced flows in liquid-filled containers and in the Earth's core," *Geophys. J. Int.* **144**, 539–554 (2001).

Richter, B., "The long period elastic behaviour of the Earth," In: *Variations in Earth Rotation*, (Eds D. D. McCarthy and W. E. Carter) Geophysical Monograph 59, IUGG Volume 9, 21–25 (1990).

Rieutord, M., "Inertial modes in the liquid core of the Earth," *Phys. Earth Planet. Inter.* **91**, 41–46 (1995).

Rieutord, M. and Valdettaro, L., "Inertial modes in a rotating spherical shell," *J. Fluid Mech.* **341**, 77–99 (1997).

Rochester, M. G., Jacobs, J. A., Smylie, D. E. and Chong, K. F., "Can precession power the geomagnetic dynamo," *Geophys. J. R. Astr. Soc.* **43**, 661–677 (1975).

Seyed-Mahmoud, B., "Elliptical instability in rotating ellipsoidal fluid shells: Applications to the Earth's fluid core," PhD Thesis, York University, Toronto, Canada (1999).

Seyed-Mahmoud, B., Henderson, G. A., and Aldridge, K. D., "A numerical model for the elliptical instability of the Earth's fluid core," *Phys. Earth Planet. Inter.* **117**, 51–61 (2000).

Smylie, D. E., Hinderer, J., Richter, B. and Ducarme, D., "The product spectra of gravity and barometric pressure in Europe," *Phys. Earth Planet. Inter.* **80**, 135–157 (1993).

Smylie, D. E., Francis, O. and Merriam, J. B., "Beyond tides – determination of core properties from superconducting gravimeter observations," *Proceedings of the 14th International Symposium on Earth Tides*, Mizusawa, Japan, August 28 to September 1 (2000).

Smylie, D. E., Francis, O. and Henderson, G., "Core properties from superconducting gravimeter data," Workshop Report, *OHP/ION Joint Symposium*, Mt. Fuji, Japan, January 21–27 (2001).

Song, X. and Richards, P. G., "Observational evidence for differential rotation of the Earth's inner core," *Nature* **382**, 221–224 (1996).

Souriau, A., "Earth's inner core: Is the rotation real?" *Science* **281**, 55–56 (1998).

Souriau, A., Roudil, P. and Moynot, B., "Inner core differential rotation: facts and artifacts," *Geophys. Res. Lett.* **24**, 2103–2106 (1997).

Stergiopoulos, S. and Aldridge, K. D., "Ringdown of inertial waves during spin-up from rest in a fluid contained in a rotating cylindrical cavity," *Phys. Fluids* **30**, 302–311 (1987).

Stevenson, D. J., "Anomalous bulk viscosity of two-phase fluids and implications for planetary interiors," *J. Geophys. Res.* **88**, 2445–2455 (1983).

Stewartson K. and Roberts, P. H., "On the motion of a liquid in a spheroidal cavity of a precessing rigid body," *J. Fluid Mech.* **17**, 1–20 (1963).

Su, W. J., Dziewonski, A. M. and Jeanloz, R., "Planet within a planet: rotation of the inner core of the Earth," *Science* **274**, 1883–1887 (1996).

Thompson, R., "Diurnal tides and shear instabilities in a rotating cylinder," *J. Fluid Mech.* **40**, 737–751 (1970).

Vanyo J. P. and Dunn, J. R., "Core precession: flow structures and energy," *G. Journal Int.* **142**, 409–425 (2000).

Vanyo J., Wilde P., Cardin P. and Olson, P., "Experiments on precessing flows in the Earth's liquid core," *G. Journal Int.* **121**, 136–142 (1995).

Vladimirov, V. A. and Tarasov, V. F., "Resonance instability of the flows with closed streamlines," In: *Laminar–Turbulent Transition* (Eds V. V. Kozlov, D. E.) Springer-Verlag, pp. 717–722 (1985).

Name index

Subject index

adaptive Gaussian kernel estimates 85–6
adiabatic state 107–10, 111
Africa 7, 9; Southern 12
Alfvén torsional waves 58, 61–9
America 11, 32; Central 43
anisotropy 102; ultrasonic wave 171
Asia 43; East 11, 20, 27, 32, 79; Southeast 22, 23
Atlantic 27
ATS (astronomical timescale) 85, 92
Australia 11, 22, 23, 27; Western 94
azimuthal flow 20, 31
azimuthal wavenumbers 181, 191, 205

BE (bulk Earth) 120
Bessel function 105
Boulder 201
Bremen 176
Brunhes–Matuyama boundary 85, 93
Brunt–Väisälä frequency 181
BSE (bulk silicate Earth) 120, 121
buoyancy forces 18, 109, 113

Cadarache 164
Carnot cycle 115
Cenozoic 90
Chandler wobble 60, 208
Chebychev polynomials 132
chemical compositional effects 44
Chile 92
Chronogram technique 84
CI chondrite 120, 121, 122
CMB (core–mantle boundary) 1–3, 13, 39, 40, 42, 43, 48, 90, 94, 104, 105, 107, 109, 111, 115; bumpy inner core surface 58; dissipated energy 181, 183, 184; electromagnetic coupling 70; gradual thermal changes 86; heat flows 6–11, 15, 18, 29, 112; models of topography 58, 72; nonaxysymmetrical topography 67–8; nonaxysymmetric geomagnetic field structure 20–7; oblateness constrained 60; partially molten regions 49; pressure exerted on 68; rotation studies 56; thermal boundary conditions 3–6; thermochemical environment near 11–12
CNS (Cretaceous Normal Superchron) 29, 78, 86, 87, 88, 91
cold material 86
conductivity 57, 62, 64; thermal 108–9
convection 2, 9, 22; dynamos and 153–79; enhanced small-scale 24; mantle 39–55, 171–6; in rotating spherical shells 130–52; turbulent 5, 149
cooling 41, 114, 118, 119
core flow 25–7
core–mantle coupling 1, 170–1
Coriolis acceleration 2, 18
Coriolis force 15, 103, 130, 143, 145, 153, 164, 166, 167
Coriolis parameter 131, 132, 168
Crete 93
cryptochrons 84, 86, 88
crystallization 109, 153, 154, 171

dendrites 171
DIP (decrease in paleointensity) 91, 93
dissipation 181; Joule 102, 103, 114, 158; modal excitation and 183–4; ohmic 100–29, 146, 148; viscous 58, 148, 182
DPIV (digital particle imaging velocity) 188
dynamos 90; convection-driven 31, 140–3, 153–79; critical 115–16; efficiency of 119; problem of reversals 148–50

earthquakes 44
Earth's core: dynamics at short periods 180–210; energy fluxes and ohmic dissipation 100–29; modelling Alfvén torsional waves inside 61–9; radioactive 120–3; response to changes in rotation of container 58–61; solid inner shape, deviations from axisymmetry of 69–70; see also CMB; inner core; outer core